The Sky's Dark Labyrinth

The Sky's Dark Labyrinth

Stuart Clark

Polygon

First published in Great Britain in 2011. This paperback edition published by Polygon, an imprint of Birlinn Ltd

Birlinn Ltd
West Newington House
10 Newington Road
Edinburgh
EH9 1QS

www.polygonbooks.co.uk

9 8 7 6 5 4 3 2 1

ISBN 978 1 84697 215 7
ebook ISBN 978 0 85790 014 2

British Library Cataloguing-in-Publication Data
A catalogue record for this book is available on request from the British Library.

Typeset by IDSUK (DataConnection) Ltd
Printed in Great Britain by Clays Ltd, St Ives plc

Content approved by NMSI Enterprises/Science Museum. Licence no. 0283.

The roads that lead man to knowledge are as wondrous as that knowledge itself.

JOHANNES KEPLER

PART I
Ascension

1
Rome, Papal States
1600

Scarlet robes were the only sure way to achieve anonymity in public. Even in the narrowest streets, people would shy away as though the garments hid a leper. When physical distance was impossible because of the crush, they lowered their eyes and scuttled past, driven by the fear of judgement. Only children would gaze openly.

Most of the other cardinals took litters so that they could enjoy looking down on their charges, and escape the worst of the summer stench, but Cardinal Bellarmine liked being among the people. Only from within the crowds could he truly feel their respect, their fear. That man in the emerald silk had the clothes of rank and privilege, but not the demeanour. With eyes darting this way and that, his garments were probably bought from the profits of short-changing his customers. Then there was the glutton leaning against the wall, still rubbing his paunch from last night's meal that could doubtless have served an entire family. And the blonde woman with tired eyes, bare shoulders and bold cleavage; her sin was clear for all to see. All of them avoided his gaze, becoming awkward and self-conscious. Such reactions convinced Bellarmine of the need for his work to continue. It gave him courage, especially on a day like today.

'I don't understand why the prisoner has remained unsentenced for so long. He was arrested seven years ago. As a heretic, he should have been burned within fifteen days,' said his companion.

'Capital punishment is a last resort, young man.'

'With respect, I'm thirty years old, hardly a young man.'

'You're half my age, Cardinal Pippe. You're a young man to me.'

They were squeezing through a passageway, knocking shoulders as they headed out of the town while the throng plodded in. The sun was not yet high enough to slice into the alley, making it a popular shortcut for those eager to escape the heat-drenched boulevards. As the pale stone walls funnelled the pedestrians together, Pippe accidentally placed his sandalled foot in the running gutter. He growled in disgust.

'But it shows weakness to prevaricate like this. Rome must be strong. In the north of Europe, I've heard that witches are burned every day.'

'We are not Lutheran barbarians with their superstitions and summary executions. Everything must have due legal process – even for a heretic,' said Bellarmine.

They turned into a wider street, the sun now fully in their eyes. It was no less of a crush, and the cardinals were still walking against the flow of people. A farmer drove an old sow patiently around them, the smell of the farmyard lingering long after the animal passed from view. Young men with flapping shirtsleeves dodged in and out, hurrying to find work for the day. Scrawny dogs followed scents, and a small girl waved the grimy air away from her nose as her mother dragged her onwards.

Old houses – survivors from the sacking of the city seventy-three years ago – lined the dirt road, which was furrowed with cart tracks and cracked for want of rain. A low cloud of dust shrouded feet and ankles.

A family had chosen that day to move home. Their belongings spewed out of the door across the dirt, slowing people down and causing much head-shaking and muttering. Their donkey flicked its tail at the buzzing flies, occasionally catching one of the passers-by. A burly man lashed another chair onto the donkey's already laden back and, in the midst of it all, the mother did her best to organise the swarming children into some kind of team.

The crack of trampled wood brought a thunderous glare from the man and an apology from someone in the crowd. The two clerics took their turn in stepping around the chattels and the children.

'Why do these people clutter their lives so?' asked Pippe, openly staring at the jumble.

'It is how they define themselves. The rich have land; the poor have knick-knacks.' Bellarmine dabbed his forehead with a lace-trimmed handkerchief. On any other day he might have been amused by the young man's annoyance.

'Shouldn't they turn to God for definition?'

'They do that too.'

They paused at a crossroads to allow a small cart to rumble by. Pulled by a slender boy, it was piled high with bolts of cheap cloth that not even the sun could brighten. Still, it was a start for him. Pippe tapped his foot impatiently, raising more dust. 'I have read that the prisoner talks of the Earth moving through the heavens.'

'We cannot burn him for that; the Church has no position on those teachings.'

'But, Cardinal Bellarmine, the Bible talks of the Sun moving across the sky.' He flung an arm upwards to the brilliant blue dome.

The older man ignored the gesture and began walking once more. 'I agree. The Sun's motion is obvious. I have talked to Father Clavius of the Jesuits . . .'

'The Jesuits.' Pippe spoke the name as if it were a curse. 'More layers of grey. Why must we continually seek their approval on matters that are so clearly black and white?'

Bellarmine glanced around the crowd, satisfying himself that no one had taken any notice of the outburst. 'We need the Jesuits,' he told the younger cardinal. 'Their missionaries are fearless. They are staunching the spread of Lutherism across Europe every day.'

'But they seem more interested in natural philosophy than theology.'

'Not all of them. But, since you mention it, natural philosophy is interwoven into our theology. It remains as it was handed down by Aristotle. The Lutherans attack us there because they think it's our weak spot, but the Jesuits can defend us; their mathematicians are without equal. Are you old enough to remember when Pope Gregory ordered ten days of October to be dropped, to bring the calendar back in line with the seasons?'

'I was twelve in 1582, of course I remember it,' said Pippe wistfully. 'How could I not remember it? My birthday falls on one of the days skipped that year. A hard lesson for a twelve year old who was left wondering if he'd have to wait another twelve months to turn thirteen.'

'Father Clavius made those calculations,' said Bellarmine. 'The old way of calculating the length of a year had thrown Easter into confusion. Now, thanks to the Jesuit method, we have the most accurate calendar in the world, and the Lutherans are still arguing about whether to swallow their pride and adopt it. The Jesuits have put us ahead.'

'And they know it. They're arrogant. The Black Pope . . .'

Bellarmine grabbed Pippe by the arm and dragged him to a nearby doorway. 'Who have you heard call him that?'

Pippe stared at Bellarmine.

Bellarmine demanded again, but Pippe did not answer.

'You refer to the head of the Jesuits as the Praepositus Generalis, never as the . . . that term,' said Bellarmine.

'But there are rumours he's going behind our backs, advising the Pope privately, rather than working with the cardinals.'

Bellarmine shook his head curtly. 'Jesuit Catholicism is not in doubt.'

'Are you afraid of them?'

Bellarmine looked away. Eventually he said, 'If the Church's hierarchy is no longer simple, it is because times demand it. The Pope will always be the head, but the Jesuits are now the backbone.'

Pippe lifted his chin. 'Well, I don't trust them.'

'Stop talking, Cardinal, before you say something that both of us will regret.' Pippe frowned, then looked directly at Bellarmine. 'You're one of them . . .'

Bellarmine nodded slowly, watching the effect of his admission. Pippe bit his lip. For a moment, it looked as if he might flee, but he controlled himself and stood his ground. 'I didn't know,' he said meekly.

'I say this as a friend, it's better to have Jesuit respect than contempt. Now let us put this conversation behind us, along with this reeking doorway.' Bellarmine cut back into the street, forcing Pippe to catch him up.

'Now, as I was saying, I've spoken to Father Clavius, and he assures me the ideas of Copernicus that Giordano Bruno advocates are unworkable. Ingenious but unworkable. They require even more mathematics than the method they're designed to replace, and their predictions for the positions of the planets are less accurate than traditional methods. The philosophers will reject Copernican ideas on those grounds alone. Of much greater concern are Giordano's comments about Christ's divinity. You have read the reports?'

'Yes, Cardinal. He believes that Christ was just a man skilled in the arts of magic.'

Bellarmine nodded. That was only the start of it. Bruno also refuted the transubstantiation of the sacrament into the blood and body of Christ, and openly denied the Virgin birth. 'His list of heresies is a long one. I'm afraid for him.'

'Afraid for him? We should be afraid for Rome. We cannot risk another Martin Luther. The world still reels from his wickedness. Half of Europe's Catholics cleaved off into Lutheran heresy because of his demonic vision.'

'That, young man, is why I have to end this business with Giordano one way or another today.'

The pair arrived at a quieter part of town. Though just a few turns from the main streets, the people had all but vanished, the hubbub dissolved in the soupy air. The calm was eerie, and Bellarmine shuddered as the gaol's oak door scraped open.

'Welcome, gentlemen, it's not often we have such distinguished visitors.' The gaoler fussed around them as if they were much loved dinner guests. Bellarmine remained silent as their host raised a flaming torch and led them down a flight of spiral steps. At a small doorway, the gaoler flicked his cape over his shoulder. 'He won't give you any trouble.'

As he unlocked the door, Bellarmine crouched to peer into the dark cell. He could make out nothing, not even the dimensions of the silent room. Only the stink of an unwashed body betrayed the presence of someone inside.

'You'll need this,' said the gaoler, handing him the torch.

Bellarmine edged inside and stood up. Pippe craned to see over his shoulder. The floor was covered in straw, and the reek grew stronger.

'Giordano, it is Robert Bellarmine.'

There was no answer. Bellarmine called again.

A coiled figure was discernible in a corner, and Bellarmine feared that he was too late, that Giordano had simply been left to rot. But the prisoner lifted a trembling hand to shield his face from the torchlight. His hair was a shoulder-length frizz of grey. His eyes were screwed shut like a newborn pup. It took a long time for them to open. Bellarmine waited. He could feel Pippe's impatience behind him but only when Bruno's eyes flickered with something that could have been recognition did Bellarmine speak again.

'I understand what you must feel.'

Giordano did not move.

'These past seven years . . . I have come to know you, to love you. I want to save you.' He offered the palm of his free hand to the prisoner, whose cracked lips began to move. His voice was a barely audible rasp.

'Fetch me water,' Bellarmine barked over his shoulder. There was a scuffling, followed by a damp wooden cup being thrust at him. He took the vessel with his free hand and held it to Bruno's lips. Pippe removed the torch from his other hand. Bruno's face was cast orange by the flames and his moving lips made words. 'Set me free . . . I will teach my works.'

'Your works are flawed. You cannot be set free until you let go of those beliefs,' said Bellarmine. 'Only Vatican theologians are permitted to interpret the Scriptures. Don't make the same mistake as the Lutherans – translating the Bible into German so that any man can draw a conclusion. It leads to confusion and collapse. You should know that.'

Bruno twitched.

'You risk death if you do not recant. Time is running out.'

'Then deliver me into God's hands.'

'Not God's hands, Giordano. You risk damnation for your beliefs.'

'The Devil will be saved too.'

'That decides it,' hissed Pippe. 'Rank heresy.'

Bellarmine lifted his hand to silence the younger cardinal.

'The Devil is beyond salvation, Giordano,' he said mildly. 'Our greatest theologians tell us so. We cannot pick and choose what we believe. Authority is handed down from one echelon to the next, with all our beliefs flowing like a river from the fountain of God. This cannot be questioned; the Church derives its strength from unity. Do not trouble yourself with matters of interpretation. How can you achieve more than the legions of cardinals who have pored over the Holy Book for centuries and refined our understanding to perfection?'

Bruno stared into the darkness.

Bellarmine felt the stone in his chest grow heavier. He was unable to keep the pleading from his voice. 'A simple recantation is all that we need from you, some act of sincere contrition. Otherwise you strike at the authority of the Church, and we cannot allow that.'

Bruno's eyes suddenly widened in the flickering light. Bellarmine's breath quickened and he leaned closer. 'Yes?'

'I have found God,' whispered Bruno. 'He is not some ethereal being but he is all around us, he is in everything . . .'

Something dark stirred in Bellarmine. 'No, Giordano, no. God stands away from his creation. You mustn't . . .'

'And the Holy Spirit? It is the very soul of the Earth beneath our feet.'

The dread grew, overwhelming Bellarmine's compassion. He locked eyes with Bruno. 'You speak of the underworld, the Kingdom of Hell. I cannot allow you to place God there, Giordano.'

'Kill me, cardinal, I invite it. I demand it. I will have the validation of knowing I was right – of taking my rightful place at God's side. When the time comes for you to kneel before me, I will be ready with my judgement, ready to send you all to Hell.'

There was no mockery in Bruno's tone. Bellarmine fought for breath. He grasped at the stone wall and hauled himself upright, sucking the fetid air into his lungs. 'You are insane, dangerously insane.' He whirled to leave the cell, colliding with Pippe and the gaoler. Sparks flew from Pippe's torch. 'There can be no earthly redemption for this man.' Bellarmine took the spiral steps two, three at a time, gathering his robes so as not to trip. His lungs screamed for fresh air. 'We have no choice but to commit him to God's mercy for final judgement.'

A fortnight later, Bellarmine awoke with the first hints of dawn. For a while he watched the ochre shafts of light cross his room. The first wafts of incense from the chapels reached his nose. He rose to wash himself but something was wrong. The water, usually so refreshing, stung his face. He stared at his hands, and into the bowl beneath them, seeing the deep lines and bulbous nose of his rippling reflection.

Then he remembered the day.

He kneeled immediately by his bed, welcoming the jolt of pain as his aching knees struck the floor. He searched himself for remorse or doubt, or anything that would need atonement.

He found none. He interlaced his bony fingers and dipped his head in prayer.

There was a sharp rap on his door. Pippe's face appeared. 'Are you coming to the piazza after Matins?'

Bellarmine swallowed, shook his head.

Pippe began to protest but a sharp look silenced him. 'As you wish.' He withdrew.

Bellarmine prayed for God to show mercy on Bruno's soul, so close now to its release.

After the morning's formal prayers, Bellarmine arrived at his office, and was assailed by memories of the first time the condemned man had been brought to him: a boyish demeanour with Dominican fringe and an endearing warmth. And a naivety that had led to this disaster. He tried to banish the images by working, but concentration was a shy visitor that morning.

Later, a dark twist of movement caught his eye. Breathing heavily, he rose to look out of the window. It was smoke, spiralling upwards from the market square. From this distance, it was impossible to tell if his request for a fast fire had been heeded. If it had, the fumes could well have overcome Bruno, sparing him the flames.

The shouts of a cheering crowd drifted on the breeze. Bellarmine's mouth dried. Pippe had been right; he should have been there at the end. In a better world, executions would be conducted in private, but, until that day, the jeering masses must bear witness, so that the warning could be carried onwards.

Goodbye, Giordano, he mouthed, *God receive you and bless you*.

Another victim of a world eager for change, where none was needed.

How many must share your fate before the Catholic Church is restored throughout Europe?

2
Prague, Bohemia

Death danced in Prague. Every hour, the tiny skeletal figure held up an hourglass and beckoned to those in the market square. He was accompanied by Vanity, in the guise of a man raising a looking-glass, Greed, depicted as a merchant shaking a purse, and an Infidel, dressed in Turkish robes.

On this particular day, as the noon bell tolled and the macabre clockwork jig played out, an astronomer stood in front of the town clock. He studied the golden icons on the face of the timepiece. Each represented the current position of a celestial object: the Sun in Libra and the Moon in Aries, edging towards full roundness.

'We arrive at a favourable time,' he said.

'Let's hurry, husband. It's damp, and these bags are heavy.'

He relieved his plump wife of a bundle and hoisted his own from the ground, then turned to the small girl beside him. 'Are you ready to move on, Regina?'

The girl carried a roll of clothing under one arm and a rag doll under the other. 'Astrid is tired, Papa. I have to carry her as well.'

'We'll soon be there,' he said, as much for his own benefit as hers. They set off across the market square, weaving through the crowds. Regina squeaked with delight at a juggler in a gaudy costume of orange and green. She held up Astrid to show her the performer. Next, a basket of dirty turnips caught her attention. She showed these to the doll too. Looking back over her shoulder, she said, 'Come along! Keep up!'

Tottering behind her were two lads from the coaching inn. For a few coins they had been eager to carry the family's trunk of essentials but they did not look too enamoured with their ten-year-old mistress.

Stallholders called from every direction, keen to sell their late harvest produce.

'Mercy! Everything is four times the price it was in Graz.'

'We'll manage, Barbara,' said the astronomer.

'How? It's already cost one hundred and twenty thaler just to move, and we still have two wagonloads of furniture back in Graz. That'll all have to be paid for once we're settled.'

He fought down his irritation, blaming his mood on the fatigue lodged firmly in his muscles. He pushed on, glancing to check that Regina was still close by. A sword-swallower had momentarily captured her attention, but she soon turned away.

Through the hoards of people and baskets, on the far side of the square, the astronomer turned into a narrow residential road.

'You want the next street for Baron Hoffman's house,' called one of the boys.

'Ah, of course. In a week's time, think how familiar these streets will all be.' He managed a wan smile, but his wife looked unimpressed.

When they arrived at the house, Barbara admired the gothic windows – each as tall as a man, arranged over three storeys – and the large arch of the entrance. She seemed to straighten up. 'It's stone, you didn't tell me that.'

'I didn't know.' Their own house had been made of draughty timber. At night it creaked, and he used to imagine that God was sending him messages.

He gathered his family and rapped on the door. As he did so, the delivery boys set down their heavy load and vanished.

The wide door swung open. A housekeeper showed the new arrivals into a panelled hallway where a boy dressed in black took the astronomer's hat, gloves and cape, and a young woman, thin as a pole, approached Barbara. 'May I take that for you, madam?'

'Thank you,' said Barbara, shrugging off her heavy travelling shawl.

Footsteps signalled Baron Hoffman's approach. He appeared from the depths of the house, a broad smile on his face. 'Johannes Kepler, we have you in Prague at last.'

Kepler, taken aback by the warmth of the welcome, clasped the outstretched arm. 'We will presume upon your hospitality only for a few days, until I can secure a place of our own.'

Hoffman waved a hand dismissively. 'Nonsense, my home is yours for as long as you need it. Any friend of Hans is most welcome. He's a shrewd judge of character and he tells me you're the finest mathematician in Christendom.'

Kepler could not help but smile at mention of their mutual acquaintance. The charismatic Bavarian Chancellor, Hans Georg Hewart von Hohenburg, was everything that Kepler admired: erudite, enquiring, gracious, well connected, high-born. He had first heard of him some years ago when a courier in bright livery arrived at the Lutheran school in Graz, where Kepler was teaching, and handed over a letter from Rome.

Standing in the courtyard, Kepler broke the wax seal and saw that it was from somebody called Father Grienberger, a Jesuit mathematician. The handwriting was composed of deliberate strokes, each character devoid of flourish, and asked Kepler whether he would help a nobleman – Hewart – with a problem of chronology.

Hewart was seeking the exact date when a magnificent constellation of stars could have appeared. The alignment was described by the classical Roman poet Lucanus in the epic work *Pharsalia*, about the civil war between Julius Caesar and Pompey the Great.

Kepler's first thought had been an uneasy one: Why were the Jesuits asking for his help? But that evening, curiosity piqued and eager to prove himself, he had brushed away his doubts along with the clutter on his desk and set to his calculations.

He first made rough guesses at the stars he thought Lucanus had been describing, then started to calculate their positions more than 1,500 years ago in the sky, compensating for the drift in the calendar, to see if they lined up. When he could find no match to Lucanus's description, he recalculated, convinced he had made a stupid error. When the answer came out the same, he tried different stars, searching for any pattern that might be reasonable. In the end, he was forced to write back to Hewart stating that the great poet had been caught out in a flash of artistic licence. No such pattern of star had ever existed in the skies above earth.

Hewart responded with more questions relating to other documents: one month it was the precise date of a conjunction between Mercury and Venus in 5 BC; another it was the date of Augustus Caesar's birth and the appropriate star chart to divine his character. Each request was designed to test the veracity of a historical document by checking its celestial descriptions against Kepler's ability to calculate the position of the stars in times past. With each answer, Hewart built a more precise chronology of history.

As the correspondence mounted, so the letters became warmer. Their sentiments transformed from politeness to respect, and gradually blossomed into friendship. Hewart would offer the young astronomer advice, and never more valuably than in recent times when a hardening of Catholic attitudes in Graz's ruling class had meant that Kepler and his family had been forced to leave because of their Lutheran beliefs.

When he heard the news, Hewart had recommended Kepler to Hoffman, an imperial advisor, who agreed to take them in. To Kepler, the act had underlined the injustice of the exile because both Hewart and Hoffman were Catholics.

'Baron Hoffman, you are most gracious to accept us on Hans's recommendation. May I introduce my wife, Barbara?'

Hoffman smoothed his chestnut hair. It was thinner than it used to be, and his doublet was a little tighter, nevertheless, he retained the power to make a woman blush just by looking at her. He bowed. 'Frau Kepler, time has not touched you.'

Her eyelids fluttered. 'Thank you, Baron.'

'Please, call me Johann. I have assigned you a maid to make your stay as comfortable as possible.' He beckoned the thin woman who had returned from stowing Barbara's shawl. 'This is Anicka.'

She bobbed at the knees. 'Madam.'

'And who is this?' asked Hoffman, crouching down.

'This is Regina, my stepdaughter,' said Kepler, placing a hand on her shoulder.

'And this is Astrid,' said Regina, offering her doll.

'A pleasure to meet you both.' Hoffman turned to Kepler. 'I have something for you.'

On the ornate hall table was a package, wrapped in waxed paper and bound with string. 'Hans was at court last week. He left this for you.'

'Thank you. How is the Chancellor?'

'In good health but rather preoccupied. I sense urgency in the diplomatic corps these days.'

Inside the wrapping was a vellum-bound book. Kepler flicked to the title page and gasped. 'Ptolemy's *Harmony* – and in the original Greek. I have coveted this for some time.'

'Hans said as much. It is to welcome you to your new life in Prague. Now, honoured guests, you must be tired. Anicka will show you to your rooms, and I will have your trunk brought up. Please join me for refreshments once you are established.'

'It will be our pleasure,' said Barbara before Kepler could reply. Once in their suite, Kepler sank into an upholstered chair, his bony body taking up only half of it. He started leafing through the book, but all too soon his eyes began to close. He was jolted back to consciousness by Barbara telling the maid where to hang dresses and shirts, how to fold stays and underpinnings, and where to place them in drawers, only to move them a moment later when she spied a better place.

'I can do all this for you, madam. You need not worry yourself,' said Anicka.

'How will I know where to find things?'

'I am your maid. You ask me, madam.' When the clothes were stored, Anicka left. All that remained in the trunk were Kepler's books and papers. They took up a good quarter of the space. 'I will sort these later,' he said and went to the window, eager for the cool air that lingered by the glass. His throat prickled.

'Are you unwell again?'

'I am starting a fever, that's all. With God's grace, it will pass.'

She tucked a lank strand of his hair behind his ear. 'You know, I don't think living in Prague will be so bad after all.'

Kepler managed a weak smile. At the next window, Regina was pointing out the sharp spires of the city's skyline to Astrid.

'Come along, you two,' said Barbara, 'we must join our host.'

Hoffman sat at a large table close to the panelled window, basking in the last rays of the afternoon sun. He stood up as the guests made their tentative entrance.

'Come in, come in. Take a seat.'

Kepler waited for Regina to hop into a chair, and then eased it into the table. He seated himself next to Barbara.

Hoffman poured three goblets and passed them round.

'To your new life in Prague,' he toasted.

The wine tasted considerably smoother than Kepler was used to drinking. Though weak, it went to work immediately, and with each sip, he felt the stiffness in his limbs ebb a little more.

'I cannot thank you enough for all you are doing,' said Kepler.

'It is the least I can do for a family who has suffered as you have. Forgive me for asking, but how bad was it in Graz?'

'Just to be Lutheran was to be a target. Every day the Archduke passed new laws against us. It was rumoured that he thought Emperor Rudolph weak because of his tolerance of Lutherans throughout the Empire. So Ferdinand was determined to set an example in his own part of it. First, our ministers were banned, then our hymns, then the possession of Lutheran books. Even to bury a child . . .' He could still feel the tiny bundle that had briefly been their first daughter, cradled in his arms. Barbara had risen on that morning and dressed in silence, then sat rocking back and forth. Kepler had seen the indescribable pain in her eyes and known that nothing he could do would erase it. Even now that impasse in their relationship troubled him on sleepless nights.

He reached over and took Barbara's hand, steeling himself to finish his sentence. 'I was fined ten thaler because I insisted on burying our child, Susanna, with Lutheran rites.'

Hoffman frowned. 'Archduke Ferdinand is pushing the boundaries of his limited authority. He knows that as Holy Roman Emperor, Rudolph cannot defend Lutherans, and so Ferdinand uses this as tacit agreement to proceed with his persecutions. It is cowardly. Someone must make a stand, but who? If Rudolph speaks up, he risks excommunication from the Vatican, and with the Empire so deeply divided right now, that would surely lead to its collapse. How did it come to this?'

'Much as it pains me to say this, there were those in our Lutheran community who invited it. Mostly teachers. They stood in front of their classes and attacked the papists like dogs slavering over old bones. In the face of their rabid insults, the Archduke found it easy to act.'

'They say he returned from Rome determined to lead his land back to Catholicism.'

Kepler nodded. 'As is his right in law. All he needed was the excuse, and the foolish teachers provided it.'

'Even so, to include you in their punishment, when you were Ferdinand's Mathematician . . .'

Kepler's body tightened at the memory of that final day. 'I served him with diligence yet, when the time came, it made no difference.'

It had been just after dawn when Kepler had taken his place among those summoned to the town church. The early hour of the call was a blessing because, the previous night, sleep had escaped him. He had prowled the house, tentacles of fear entangling his insides.

Eventually settling into an exhausted heap at Barbara's feet, he crossed his arms on her lap and rested his head. She closed the outlawed prayer book and ran her fingers through his hair. Together they had waited for sunrise, and judgement.

By six in the morning, the crowd was a thousand strong. As the numbers rose, so did the heat. Kepler edged onto a worn pew as officials placed the city rolls on a table in the middle of the church. Behind them was an elaborate wooden throne on a raised dais. Near the altar, a black-robed priest was swinging a smoking thurible of incense, blessing anything within reach.

'Courage,' whispered a passing acquaintance, jarring Kepler from his thoughts. Another squeezed his arm, undisguised pity on his face and possibly a hint of shame. He's going to convert, thought Kepler, experiencing a stab of betrayal.

His tension rose with the clatter of horses' hooves outside. The congregation stood and waited for the procession to appear, all eyes on the young Archduke.

Although twenty-two, Ferdinand still looked as if he were a boy dressed to resemble a man. There was no definition in his doughy cheeks, just a long nose that slid down his face. His sandy mop bounced in time with his skittish gait and his thin moustache had been waxed and kinked upwards. He wore a partial suit of black armour, ludicrously teaming it with riding boots of pale brown leather and a wide-brimmed felt hat.

A phalanx of guards clanked around him, their armour polished like mirrors. Behind them, more officials walked with exaggerated gravitas. These were the commissioners who would examine each member of the congregation and decide their fate.

Kepler watched as the Archduke advanced, but the young ruler stared ahead with practised aloofness. As he passed, Kepler despaired. Only then did he realise that he was still harbouring the faint hope of recognition and reprieve.

Ferdinand sat on the makeshift throne and signalled for the proceedings to begin. During the inflammatory sermon that followed, the Catholic preacher hurled back all that the Lutherans had dished out. Kepler shut his eyes and silently muttered a prayer, calling for strength and begging forgiveness for the pain he was about to inflict on his family.

One by one, the men were called to the central table where each professed their obedience to Rome – even those who a week ago had been screaming insults at the Pope. All was apparently expunged by this public conversion. When Kepler's name was called from the register, he rose from the pew and walked towards the table, legs unsteady and blood pounding at his temples.

The commissioners regarded him with graven faces. 'Johannes Kepler, you have been called here today so that we may examine your faith. Do you understand?'

'I do.' He searched each face at the table.

'Do you worship in the Roman Catholic way, with your trust placed in God through His Holiness?'

Kepler spoke clearly: 'No.'

A murmur of excitement rippled through the crowd.

'Are you willing to swear your allegiance to Rome?'

'No.'

There was a collective gasp from the crowd. The commissioners held a hushed discussion involving much head shaking. Eventually the central official stood up. 'Johannes Kepler, you are a heretic. You and your family must leave Graz and the entire territory of Styria within six weeks. If you return, you do so on pain of death.'

Kepler looked to the Archduke, who rolled his eyes as if bored.

Hoffman blew out a long breath as Kepler finished his story. 'What a thing to have to endure. Take comfort in knowing that this cannot happen in Prague.'

'I'll be honest with you, I fear that the tolerance that once gave our Empire unity is slipping away. Are we not all imperial subjects regardless of personal belief? I wonder if anywhere, other than the Lutheran heartland to the west, is safe for my family now.'

Hoffman waved the objection away. 'Fear not, Johannes. We cannot be judged by what happened in Graz. Rudolph may be sworn to Rome but – like his father – he is a tolerant man. You and your family are safe now.'

3

Kepler yelped like a puppy as the broom cracked him on the ankle.

'Sorry,' mumbled the boy who was clumsily brushing away the hair from around Kepler's seat.

The barber cuffed the lad, who retreated to a safer corner of the shop, then reached for a sheet of polished metal. 'There, sir, all done.' He held the scratched mirror for Kepler to peer at himself. Massaging his bruise, Kepler stared incredulously. He did not know whether to be fascinated or appalled; he hardly recognised himself any more. He had long become accustomed to the scars from his bout of childhood smallpox, but in the six months since Susanna's brief attempt at life, the flesh of his face had fallen away. Although more than a year shy of his thirtieth birthday, his once plump cheeks had sunken into dark wells and his eyes had grown heavy cowls. While the barber had done a good job of restoring some vigour to his hair, it now seemed in conflict with his cheerless face.

He paid for his haircut and left, eager to reach his next appointment. As he walked the overcast streets, he wondered what others thought when they looked at him. *Perhaps they think me close to death.* The notion gave him some curious satisfaction. A grand funeral would finally show the doubters what a valued citizen he had been. He conjured pictures of mourning crowds, the Archduke's arrival in black, and the narration of condolence letters from universities across Europe.

'Johannes!' The call broke him from his grim musings. 'We're over here.'

Hoffman was standing with a stocky man in a short cape. The stranger favoured Kepler with a polite smile.

'Allow me to introduce Jan Jessenius, anatomist and fellow advisor to our illustrious Emperor Rudolph II,' said Hoffman.

'An honour to meet you, Herr Kepler. I have heard of your great book, though, alas, I have yet to experience the pleasure of reading your *Mysterium Cosmographicum.*' His words were warm, but his eyes scrutinised Kepler.

'The pleasure is mine, Herr Jessenius.'

'Gentleman, I thought we would take lunch at the inn.' Hoffman indicated a squat building with a black timber frame, leaning, as if inebriated, on its neighbour for support.

'Lead on,' said Jessenius, clapping his hands and rubbing them together.

Inside, they burrowed through the drinkers and searched for a spare table. As they squatted on stools around a battered wooden slab, the conviviality of the place lifted Kepler's spirits. It was as if the inn were divorced from the troubles of the outside world.

Hoffman leaned towards him. 'Jan performed the first public dissection of a human cadaver here in the city, not three weeks ago. If I had not seen the lengths of the intestine myself, I would not have believed it.'

'Thank goodness the day was a cold one – it kept the stench at bay,' Jessenius added, prompting the men to grin.

'I have considered turning to anatomy myself,' said Kepler. 'I find the resonances between the condition of the individual and the aspects of the heavens fascinating.'

As any learned man knew, the twelve constellations of the zodiac influenced the twelve regions of the body, starting with Aries and the head, face and brain, Taurus and the neck, throat and larynx, and continuing downwards to Pisces and the feet and toes. As the planets passed through these constellations, so they exerted their own power on those areas of the human body: Jupiter brought with it a desire to hunt; Saturn, the blight of melancholia; while Venus stirred the passions. The planets and their ever shifting alignments combined to produce the celestial wind that swayed the human soul, sometimes provoking insight and brilliance, happiness and strength; at others despair and illness, even wickedness if the alignment were adverse.

'But,' continued Kepler, 'my heart is drawn so powerfully to the stars that I cannot help but think astronomy is God's choice for me.'

A barmaid placed a pie and three pewter plates on the table. Hoffman unsheathed his dagger and levered open the pie's coffin lid. He and Jessenius both helped themselves to large platefuls of the pink and brown meat, whereas Kepler picked more carefully.

'I believe that you chose Prague for your exile with an agenda in mind,' said Jessenius.

'Indeed. I hope to work with the Danish astronomer Tycho Brahe. He has recently taken up residence not far from here, at Benátky Castle.'

'Why Tycho?'

'He's the greatest observational astronomer alive. He has spent his lifetime on the subject and amassed the finest collection of observations in human history.'

'Forgive me,' said Jessenius, toying with some meat, 'but I thought your allegiance was to the Bear, Ursus, the imperial mathematician.'

Kepler felt his cheeks burn. Would this haunt him for ever?

Hoffman looked concerned.

'I did not give Ursus permission to publish my letter as the prelude to his book,' said Kepler.

'But you did write the letter.' A watchfulness returned to Jessenius's eyes.

Hot sparks flashed inside Kepler. 'And how I regret it now. I was young, guileless and stupid, trying to worm my way into the favour of any astronomer who would give me credit. So, yes, a few years ago I sent Ursus a letter of the utmost praise with a copy of my book; as I did to Tycho; as I did to many others. At the time, I was unaware of their rivalry. Now, I would not hesitate to name Tycho the greater astronomer. He knows this; we have corresponded since. He even praised my work, inviting me to Prague so that we could discuss astronomy. But I was not at liberty to leave my teaching post.' Kepler scratched his head, desperate to make them believe him. 'You see the *Mysterium* is just the beginning. It sets out my belief in Copernican astronomy but not my proof of it. It is infatuation without marriage; a man's ideas without the womanly curves of a solution. Just as a carpenter needs wood to fashion, so a mathematician needs numbers to shape. In his lifetime Tycho has accumulated more observations than all other astronomers in history put together. With them, I can prove that the Sun is the centre of the Universe – I know it.' He lowered his voice to a whisper. 'I feel the influence of God in my belief.'

A sideways glance passed between Hoffman and Jessenius. When he realised that Kepler had seen the look, Hoffman hastily slid the pie across. 'Do eat, Johannes, or there will be none left.'

'I've had my fill, thank you,' Kepler said pointedly, jumping up. 'I've also written to my old tutor at Tübingen, asking about a professorship there. Perhaps coming to Prague was a mistake.'

'Patience, friend,' Jessenius said, wiping his lips with a napkin.

Hoffman placed a hand on Kepler's arm, guiding him back into the seat.

Jessenius continued. 'Please accept my apologies for questioning you, but you must understand that Tycho is a law unto himself. If you are to work with him, your loyalty must be beyond reproach.'

Kepler looked from Jessenius to Hoffman, and saw in their faces the reality of the situation.

'Jan is a close friend of Tycho,' confirmed Hoffman.

'Fear not, Johannes,' said Jessenius, 'I believe you are a good man and I will take word of your arrival to Tycho.'

'Sir, may I humbly beg that you not reveal the circumstances of my arrival? You see, it is my hope to collaborate with Tycho as an equal: his observations and my mathematics. Any hint of my plight will make me seem in need of charity. It would destroy my pride.' If not for the table, Kepler would have sunk to his knees.

Jessenius nodded. 'I will inform him only of your arrival. The rest is down to Tycho. Be warned, Johannes, no man makes up Tycho's mind for him.'

4

'That'll do,' said Barbara. 'Do you want me to stop breathing?'

Anicka tied the laces of her mistress's stays and helped her into a dark blue dress that flared at the hips and had an embroidered bodice.

'And the ruff,' said Barbara, oblivious to the maid's smirks as she secured the cartwheel of fabric.

'Come, husband, we should be downstairs by now.'

Kepler reluctantly put down *Harmony* and blew out his reading flame. He ran a cursory hand through his hair.

'Better than that.' Barbara pointed to the brushes on the mantelpiece. Kepler smoothed his hair backwards with the wiry implements. 'There, am I presentable?' He was wearing his best black jacket, as befitted a formal occasion, to which Barbara had sewn lace cuffs that picked out the white of his new hose.

'You'll do,' she said.

They made their way downstairs, drawn by voices and music, to where Baron Hoffman's grand reception room was already full of visitors. Everything sparkled in the candlelight: the wine glasses, the jewellery and the men's buttons. Gentlemen were in earnest discussion. The women were nodding politely or clustering in little groups of their own to exchange confidences. In the corner of the room, a quartet of musicians plucked and blew their way through a selection of airy melodies.

Their host met them at the door. 'Welcome to the Feast of the Hunters' Moon.'

'What better omen for an astronomer's first weekend in the new city?' smiled Kepler. The lively babble of conversation enveloped them. Almost at once, men eager to be introduced to the new arrival besieged Kepler. He was driven deeper into the room, leaving Barbara stranded. Self-consciously she scanned the assembly. The women were wearing high collars that plunged downwards to the swell of their *décolletages*. A further glance around the room confirmed the ubiquity of the fashion; each woman was revealing skin, in fact flaunting it.

'You must be Mrs Stargazer,' said one of the guests, older than Barbara but taller and slimmer.

'Barbara Kepler, madam.'

'I hear that your husband is a clever man. His arrival is the talk of the town.'

Barbara stopped short. 'Really?'

'Oh, yes, another for Rudolph's inner circle, no doubt.'

'The Emperor?'

'He collects thinkers the way a small boy hunts for spiders. I say! Is that what they are wearing in Graz these days?' She favoured Barbara with an unnerving smile. 'I haven't seen such a ruff in Prague for years. Still, it's good to know the old ideas live on in other places.'

Barbara touched the starched fabric standing proud of her neck by some four inches, each point culminating in a bead. She forced herself to laugh as though she had been caught in a moment of forgetfulness. 'I am unused to Prague's customs, having only just arrived.'

'Oh, my dear, a little rustic charm is welcome. It reminds us who we are.' Again that smile flashed.

'Would you excuse me for a moment?' Barbara ignored her companion's puzzled expression and retreated to the quiet hall, all but tearing the ruff from her neck. Flushed with embarrassment, she was about to thrust the offending garment behind a chair cushion when another thought struck her. She loosened the drawstring on her chemise and tugged down the neckline as much as she dared. Then she turned the ruff around, so that the opening was in front of her throat and tucked the open ends underneath the shoulders of her bodice, forcing the ruff to stand up like a collar. Catching her faint reflection in a window-pane, she squared her shoulders and returned to the party.

She spied her acquaintance and walked straight up to her. 'I'm back,' said Barbara.

She received a cold look at first but watched it transform into surprise and then warmth as the older woman registered Barbara's altered appearance. 'I am Frau Dietrich. Now, let me introduce you to my friends.'

In the gentlemen's quarter, Kepler sipped a rich wine, delighted to be drinking from glass rather than pewter. The cut crystal felt so much cleaner on his lips. One day, he thought, Barbara and I will have a small set just like these.

'My dear friend, I trust the book is to your liking.'

There was no mistaking the voice of Hans Georg Hewart von Hohenburg.

'Hans, how good to see you again,' said Kepler.

The Bavarian Chancellor was a short man, no taller than Kepler, but carried himself much straighter to give the illusion of height. As always, he was in the best of clothes; this evening clad in an exquisite jacket in maroon velvet and brilliant white hose. Every blond hair on his head had been brushed strictly into place. But when Hewart thought no one was looking, he was in the habit of sucking on his bottom lip, as if chewing over some conundrum.

Kepler noticed how delicately Hewart held his glass by the stem, and readjusted his own ham-fisted grip. 'The book is a revelation and an inspiration all in one. I can scarcely set it down. And what of the book I recommended to you?'

Hewart smiled. 'Ah . . . I'm afraid I've let you down. I can understand so little of what Copernicus writes that I've given up. He puts in so many epicycles that I just cannot picture the convoluted motions – all the planets whirling through the heavens so. To my thinking, it is scarce improvement on Ptolemy.'

'Copernicus over-complicates his system but his basic idea is sound.'

'Can the Sun truly be at the centre of everything?'

Kepler drew closer. 'A growing number of us think so. There is an astronomer in Italy named Galileo . . .'

'A Catholic?' Hewart's voice rang with pride.

'An astronomer,' said Kepler by way of refusing the distinction. 'He writes to me, signing himself one Copernican to another. Yet he will not speak out in favour of the system as I've urged him to do.'

Hewart tugged at his goatee. 'What holds him back?'

'We cannot yet prove that Earth moves. Until we can, I fear there will always be support for the old ideas.'

'But how could you measure such a thing?'

'Simply. If the Earth orbits the Sun, the North Star will appear to move during the year.'

Hewart looked at him blankly.

'Here, hold your finger in front of your face and close one eye. Where's your finger against the background?'

'In front of the lute player.'

'Now look through the other eye.'

'It's moved; it is to the left of him now, but I haven't moved my finger.' He swapped eyes again, testing the new discovery.

'Precisely, it's called parallax. Your finger has remained stationary but it appears to have moved against the more distant objects because you have changed your vantage point. The same will happen to Earth. Every six months we look out from the other side of our orbit. All someone needs to do is measure the position of the North Star at six-monthly intervals and see it change.'

'You say that Copernicus complicates his system . . . Can it be simplified?' Hewart asked.

Kepler nodded, setting off miniature waves in his wine. 'The key is beauty. Ptolemy and Copernicus both present ugly systems of motion that no man can keep in his head. The true movement of the planets will be a simple elegant dance – beautiful even. How could the heavens be otherwise?'

'You look flushed. Has the wine disagreed with you?'

'I am fighting a poor humour, that is all.' Kepler raised his free hand. His forehead was clammy again. His vision started to blur. 'We travelled through so much country on the way here, there is no telling what miasmas we encountered.'

'My dear friend, let us sit you down.' But Hewart's eyes were drawn away.

Kepler followed the gaze and saw Jessenius approaching. 'I'll be alright.' He took a deep breath and pulled himself straighter. Alongside Jessenius was a young man, tall and well built, with intense pale eyes and a walk that bordered on a swagger. He was dressed in green velvet with black hose and held steady a long sword, sheathed at his side.

'Johannes, allow me introduce the Junker Franz Tengnagel, one of Tycho's assistants. He brings word from Benátky Castle.'

'Herr Kepler,' said the young man in clipped tones, thrusting forth a sealed letter.

Hesitating at first, Kepler took the letter and slipped a finger under the seal. Blinking to clear his vision, it was difficult to read Tycho's extravagant handwriting despite the mass of candles that poured light around the room.

'Well, let us share in your news,' Hewart encouraged.

Kepler skimmed the words again, took a deep breath. 'I am welcome to be his companion in observing the heavens.'

'Then we recharge our glasses,' said Hewart, 'and drink in your honour.'

'We ride to Benátky the day after tomorrow,' said Tengnagel. It sounded like an order.

After the toast, Kepler slipped out to read the letter more carefully. His surroundings had begun to assume an unreal edge, as if he were looking at them through old panes of glass. Yet Tycho's words were imprinted on his mind. *You will come not so much as a guest but as a very welcome friend and highly desirable participant and companion in our observations of the heavens.*

Kepler's cheeks became suddenly damp with tears, and his body began to tremble. He leaned back against the wall, feeling the corner of an ornate mirror-frame press into his back.

When the peculiar exorcism had run it course, he wiped his eyes and pushed himself away from the wall. His mind was clear. The hallway looked normal again. Then a labouring voice caught his attention. It called his name, though more in statement than in greeting. A large man was tottering near the staircase. He was bound into a suit of silver-grey cloth, rolls of fat bulging between the strapping that held the seams closed. His bald head emerged from a ring of blubber.

'I am Nicholas Reimers Ursus, Mathematicus to his Imperial Majesty, Emperor Rudolph. You perhaps know me best as The Bear.'

Kepler caught his breath. Ursus, The Bear. 'You caused me trouble, sir, publishing my private letter as if I sided with you against Tycho Brahe, the prince of astronomers.'

Ursus snorted, seemingly amused. 'The prince of astronomers, you say? I worked as one of your "prince's" subjects, just as I hear you are about to do. Be warned, he is not what you think.'

'He has gathered the finest astronomical observations in the history of mankind.'

'For all his work, Tycho is an anchor to progress. Who cares about his arrangement of the planets – or my one? Both are wrong. You know that as well as I do.'

Kepler nodded cautiously.

The Bear squeezed out his words between ragged gasps. 'I'm under no illusion about my worth as an astronomer. I'm no great asset to history. Neither is Tycho. He will squander the measurements, if you let him, and the new thinking will never come. You're his best hope for immortality. Examine everything you thought you knew; leave no assumption unchallenged. If it cannot be proved, it can be changed. I'm too old to put this insight to use but you . . . you are different from any man I have encountered before. You are original. Your *Mysterium* proves that. I published your letter without your consent – that is true – but not to cause you trouble. It was so that those yet to come will see the praise you once lavished on me.'

With a grunt of exertion, The Bear lumbered into motion, headed for the front door and, all too soon, was gone into the night. Only then did Kepler's mind fill with the questions he should have asked.

Turning back towards the party, he all but collided with the loitering Tengnagel. 'Do excuse me, Junker Tengnagel.'

'You are acquainted with The Bear, Herr Kepler.'

'I am not. I've simply had dealings with him.' Kepler walked on, sensing the other man's eyes following him back into the reception.

It was long past midnight when Kepler followed his wife to the bedroom. Downstairs, a few guests still lingered, seemingly content to see the celebration through to the dawn.

'You are showing much courage tonight, wife.' Kepler planted kisses on her exposed neck and chest. Her skin tasted sweet to his wine-moistened lips.

She pushed herself further into his caress. 'It's the fashion. And I need one of those tapered bodices – all the ladies were wearing them tonight.'

'I didn't notice,' he said.

'Really? They were so beautiful.'

'There is only one star in my Heaven.' He kissed her firmly on the lips, thrilled by the hunger with which she responded. He pressed her to the bed, the alcohol accentuating the dizziness of their fall.

'Think of how we got here,' he said. 'I would not have thought it so sweet to suffer the injury and indignity of being forced to abandon house, fields, friends and homeland for religious belief. If this is also the way with real martyrdom, how much bigger the exultation must be to actually die for one's faith.'

'Oh, husband, you do choose your moments to say such funny things.'

'Then I will stop talking.' Kepler let his hands rove across the swell of her torso, finding the laces on her dress.

'Oh, I can't wait for you to undo all of those.' Giggling, she grabbed at her skirts and petticoats, and bundled them upwards, covering her face.

Later, she lay gazing at the ornate ceiling; her arms swept back on the pillow, her breathing deep and contented. 'So many people of such high standing,' she mused. 'Once we look the part, we will have no problem finding you a job among them.'

Kepler laughed beside her. Shading his eyes with his hand, he pretended to look into the distance. 'It is to noble Tycho that I fix my gaze. He is my prince, and at his side will I find my station. Once I am established, I will send for you. In the meantime, the Baron has agreed you can stay here.'

'Yes, but if it doesn't work out at Tycho's . . .'

Kepler silenced her with a gentle kiss. 'It will.'

5

Kepler clung to the reins with grim determination, wincing every time the saddle jarred his bony posterior. Ahead of him, Tengnagel crouched low on a white stallion, driving headlong into the winding approach road to the castle. Kepler caught a glimpse of the whitewashed stone perched high above Benátky village before the cottage-lined street filled his vision and the dwellings reverberated with the sound of their gallop.

By the time the castle gates hove into view, Kepler was more than a dozen lengths behind, his head whirling from the strenuous ride. When he finally reached the courtyard, Tengnagel had already dismounted and was strutting about, leaving the stable-hands to calm the stallion.

Kepler drew his horse to a stop and slumped along its neck. His heart threatened to burst from exertion, and despite the autumnal chill he was sweating from every pore. He slid to the ground, his knees buckling upon contact.

A gruff voice rang out across the yard. 'What have you done to our guest?'

'I cannot be held responsible if he is unused to our pace of life, sir,' replied Tengnagel.

'I asked you to escort him here as an honoured guest, not ride him into the ground, you fool.'

Tengnagel flinched. Throwing his riding gloves into the mud, he stamped inside.

'What does my daughter see in him?' the gruff voice muttered.

Kepler rubbed his watering eyes and brought them to bear on the source of the voice. A mockery of a human face looked back. It wore a hat reminiscent of a Turkish cupola. Beneath it, two rheumy eyes of hazel gazed unblinkingly from deep sockets. A tightly rolled moustache framed a gappy mouthful of saffron yellow-stained teeth. But that was not the worst of it. The bridge of the nose had been replaced by a cylinder of rose-coloured metal, and gobs of some thick unguent clung to the margins where angry flesh met the hideous prosthetic.

Could this be the great Tycho Brahe?

The face spoke. 'I must apologise, more than I had intended, for not escorting you from Prague in person.'

He was so old. The picture in the frontispiece of his book painted him as a young noble, strapping and brave, not as this ugly palimpsest. Disappointment mingled with Kepler's physical discomfort, and he thought for a moment he might vomit. Gulping down the nausea, he said, 'I'm here now.'

A flicker passed across the old man's wide face.

I should have been gracious, thought Kepler, his head pounding.

Tycho began a slow waddle to the stone entrance with Kepler in his wake. The courtyard was strewn with timber and trestles. Two carpenters in heavy linen smocks tugged a saw back and forth between them, cutting a tree trunk into beams.

'You join us at an exciting time,' said his host. 'We are nearly settled into our new home, and the observatory is close to completion.'

Inside the castle, all was noise and motion. Giant sheets hung where craftsmen were working the stone, and Tycho batted his arms at the puffs of dust that seeped through the gaps in the coverings.

'Let me take you straight to the observatory.' He guided Kepler past wooden scaffolding and under large beams supporting the ceilings to a wide staircase where four servants sweated under the weight of a steel framework. Each side of the square contraption was larger than a man. Inside was suspended a curving track along which ran a moveable armature.

'Heavens above, that's a quadrant,' said Kepler.

'Set that down,' Tycho barked at the servants.

They lowered their burden with a clank onto one of the stone steps and retreated to the balustrades.

'Take a look.' Tycho gestured towards the object.

Marvelling at the engraved scale, Kepler ran his fingers across the cool metal. 'I've only seen hand-held ones before.'

'You perhaps now glimpse what I have achieved.'

'It was never in doubt, though I admit the grand scale of it astounds me.'

'Then let me astound you even more.' Tycho beckoned the servants back to their task and continued to climb, heaving himself up one step at a time.

Upon reaching one of the upper landings, Kepler stopped in his tracks. A miniature person was writhing on the floor, stubby limbs scratching at the air. As the bundle of flesh and clothing struggled, it babbled inanely in a squeaky voice.

'Allow me to introduce Jepp, my constant companion and castle seer.' Tycho kicked the collection of tiny arms and legs. It sprang to its knees, spitting and clawing the air in Kepler's direction, forcing him to step back.

'Evil walks today,' said Jepp, the words rising with perfect clarity from his incessant mumbling.

'What does he mean?'

Tycho laughed with a shrug. 'He is a harmless dwarf. Come, let us leave him to his ravings.'

Unnerved by the way the twitching face stared at him, Kepler broke eye-contact and hurried after Tycho. They climbed one further flight of stairs and emerged onto the castle roof. A cool breeze offered balm to Kepler's cheeks.

At first he thought his eyes were playing up again; magnifying the world to trick him into thinking that a giant had laid down a set of astronomical instruments for his miniature human helpers to tend. But no, this was reality. Tycho Brahe had created a marvel.

Kepler gawked at the towering structures of wood and metal jammed onto the rooftop so tightly that there was hardly any room to move between them. He revelled in the geometry of the instruments with their metalwork of triangles, squares, circles and spheres. The shapes were the very embodiment of the mathematical arts, the essential interface between man and the cosmos. By lining them up with the stars and planets, they could provide everything – angles, altitudes, azimuths – all the measurements that Kepler needed. They were instruments of divine astronomical purpose. This was more than an observatory; it was a shrine to the universe, with Bohemia stretching out below in a patchwork, reaching from one village to the next.

A blond man of regal bearing was cradling a compass and squinting through the sightline of an upright circular frame. As Kepler watched, the man touched the structure as gently as if it were his lover's cheek, nudging it imperceptibly.

'This is Christian Longomontanus. I lured him here to help in our new quest, though he is homesick for Denmark. Tell him, Christian, there is no better observatory – nor master – to work for.'

'All that you say is true, my lord.' He spoke in a deep voice with measured words.

'This is Johannes Kepler.'

Kepler nodded in greeting.

'A pleasure to meet you, Herr Kepler. Your reputation precedes you.'

Tycho quickly gestured to the circle. 'How is it?'

'North–south alignment is finished. We can complete the equatorial alignment tonight, if the weather holds.' He glanced up at the whitening sky.

'This is an armillary sphere, is it not?' asked Kepler.

'Yes, but stripped to its bare essentials; no need for all those other great circles. The weight flexed the metal and ruined the accuracy,' replied Tycho.

'We can measure stellar positions to better than an arcminute with this,' Longomontanus added.

'And that's not the best.' Tycho spoke with the enthusiasm of a parent. 'The wooden sextant over there can measure to thirty-two arcseconds.'

'Arc*seconds*?'

'Indeed.'

Numbers lined up in Kepler's brain. 'Thirty-two arcseconds is less than two hundredths of the full Moon's width. Your observations are nearly twenty times more accurate than Copernicus worked with. You are . . . you are beyond anything I ever imagined.'

'That is why Copernicus was wrong, and I am right.'

'Have you seen parallax?' asked Kepler.

'Never. Not even with these perfect instruments. The Earth does not move.'

'But it must!'

'I believe only what my eyes and instruments tell me. You would do well to do the same, Johannes. Now, enough of astronomy. I will have you escorted to your room, so you may rest. Then, I will see you for dinner.' Tycho clasped the expanse of his own stomach. 'We eat at three o'clock, so that our food is well digested before the night's observing begins. You look as if you could do with some fattening up.'

Kepler's saddlebags lay beneath the window in his room. He thought briefly about unpacking the various items he had brought – mostly books and papers – but at sight of the bed, he rolled onto the straw mattress. He wondered briefly what Barbara and Regina were doing back in Prague, before losing himself to a dreamless sleep.

He awoke with a start at the sound of the door.

'Who's there?' he said to the intruder.

'Herr Kepler, forgive me for waking you but the assistants all share rooms at Benátky.' Longomontanus averted his eyes.

It occurred to Kepler that he must look ridiculous, sprawled in his clothes in the afternoon. He was hot and his throat burned.

'And, sir,' said Longomontanus, pointing with his long fingers, '*that* is your bed.'

In the dusty far corner was a smaller cot, topped with greying blankets and pillows.

Kepler found Tycho in an antechamber, surrounded by assistants. Tengnagel hovered near the periphery, chin in the air, nodding enthusiastically whenever the Master spoke in his direction but at other times letting his eyes wander.

Kepler's hurried footsteps drew their attention.

'My lodgings are unacceptable,' he declared. 'I have a wife and stepdaughter, we cannot be expected to share . . .'

Tycho lifted his hand. 'Did you smuggle them in with your packing?'

Tengnagel guffawed. The others swapped sidelong glances. Regardless, it inflamed Kepler more.

'They will be here with me. I will need somewhere quiet – and uninterrupted time to perform my calculations.'

Tycho reached into a pocket and removed a snuffbox. He tipped the lid and dipped a finger into the waxy substance inside, then smeared it around his metal nose. With his hand in front of his face, he mumbled, 'And you will get them.'

Heads turned towards the Master.

'Come, let us eat.' Tycho lurched into motion. The assistants shrugged to each other and eyed Kepler, who dropped to the rear of the group. Tengnagel barged past to take up a position at Tycho's side.

The dining hall was still being set when the entourage swept into the room, sending the servants into a frenzy. Their activity set the wall hangings swaying, bringing a strange animation to the mythological depictions.

The tables were arranged as three sides of a square. Tycho indicated a chair on the left-hand table, nearly fifteen places away from the ornate seat at the centre of the top table. 'Herr Kepler, please be seated.'

The other assistants were taking their positions nearer the top table, and Tengnagel rattled a chair within three of the central seat. Kepler looked up to query the placement, but there was a warning in Tycho's eyes.

'You are most kind,' said Kepler, shrinking into his place. He watched sullenly as Tycho's wife and eldest daughter took their places at the top table, exchanging greetings with the observatory assistants. Around them clustered some of his host's other sons and daughters, each displaying the Tychonic red hair. When late guests hurried in to take seats next to Kepler, he favoured them with only the briefest of acknowledgements.

The room hushed as Tengnagel stood to say grace. It proved to be a ponderous monologue.

'Oh, do hurry up, my insides are screaming with hunger,' Tycho said during a particularly grandiose passage.

The room roared approval. Only Longomontanus continued to pray, Kepler noticed.

The servants appeared and piled plate after plate onto the long tables. Kepler watched Tycho hoist a roasted woodpigeon from a platter and rip it to pieces before chucking the carcass onto the floor. *No, not the floor!*

Underneath the table, dressed in the garb of a court jester, Jepp crouched at his Master's toes and feasted on the leftovers. When waiting for another carcass to fall, he shook the bells on his costumes in some childish rhythm.

'Ooh, do try the tansy,' urged one of Kepler's neighbours, indicating an omelette with edges as grey and ragged as old lace.

Kepler took some just to appear polite. He had always been of the opinion that food exerted a powerful influence over its consumer. If something looked bad from the outside, it probably did something bad on the inside too, but the tansy pleasantly surprised him. It tasted much better than it looked, so he took a little more. Soon afterwards another course was paraded around the tables: a predictable boar's head made up the centrepiece and was placed before Tycho.

'What is the special occasion?' asked Kepler.

His neighbour laughed raucously, jowls quivering and spilling his wine. 'Special occasion? Nothing. You're at Tycho Brahe's now.'

Kepler chewed on, picking his morsels carefully, his mood unimproved. At the top table, Tycho conversed and laughed with his assistants.

As the meal entered its second hour, a dark form banged on the table in front of Kepler, bouncing the tableware. An unmistakable jingling followed.

'Does the Master's food not suit your genteel palate?' Jepp was standing astride the platters, bulbous head cocked. His piggy eyes were fogged and his breath stank of wine.

The room was suddenly quiet. Kepler knew Tycho was watching, a leer on his greasy lips.

'Are you used to finer things in *Graaaaaz*?' Jepp drew out the last word into a song of ridicule.

'Forgive me, I am not myself today,' said Kepler, addressing the top table.

'Who are you then?' Jepp squeaked. 'Copernicus perhaps?'

'Enough Jepp, leave him be,' called Tycho.

After a moment, the dwarf's posture relaxed and he retreated to the edge of the table. But at the last moment, he lunged back in Kepler's direction. Instinctively Kepler pushed himself away. The rear legs of his chair caught on the lip of a flagstone, and Kepler tipped over, cracking his head. Jepp perched on the edge of the table, watching his victim.

The guests roared with laughter. Jepp somersaulted from the table into the middle of the room and bowed, drawing more howls of delight from the onlookers. Shaking with humiliation, Kepler turned towards his host. The great Tycho Brahe was looking back, roaring with laughter.

*

Each evening, the assistants met to discuss the coming night's work. They stood in a huddle and listened as Tycho informed them of their priorities and the division of labour. Once the programme of work was clear to everyone, they wrapped themselves in heavy capes and set off up the staircase. Their robes bestowed the illusion of priests ascending to worship the heavens.

Kepler stayed on the outskirts of the discussion and was assigned to help Longomontanus. 'I had not anticipated taking part in the observing,' he whispered to his room-mate. 'I'm inadequately clothed.'

'I have a spare cloak you may borrow.' They detached themselves from the procession and headed for their room. Once there, Longomontanus opened a cupboard and handed over a musty-smelling garment. Kepler swung it around himself. A clear foot of material pooled on the floor.

'You will be the warmest of us all,' grinned Longomontanus.

When they arrived on the roof, the great nocturnal beast of the observatory was stirring into action. A dozen shadowy figures moved between the silhouettes of the instruments, preparing them for the night's observing. During the day, the devices had been clamped rigid; now they were set free with the turn of fist-sized screws. Agog, Kepler watched the shadowplay; it was as beautiful as a dance. The operators merged with their mechanisms, each contributing to the choreography. An armature glided up a curving frame; a semicircular framework rotated into place; a triangular chassis tilted like an eagle catching an updraft. He followed the line of the instrument upwards, marvelling at the glittering stars.

Kepler thought briefly of his own hopeless attempts at coaxing the sky out of its secrets back in Graz. He had built a mound of earth on which to rest a lashed-up cross-staff of wood – and he had dreamed of measuring the parallax like that. What a fool he had been.

Below the insulation of the night sky, all dreams seemed real but all fears were magnified too. 'I have something else to tell you,' he said to Longomontanus's shadowy face. 'I'm not a good observer. I have tried, but my eyes are weak; the result of smallpox when I was a child.'

Longomontanus handed him the observing logbook. 'Then you will be my amanuensis, and no one need know.'

The Dane unlocked the giant sextant and swung it towards Deneb, their reference star for the first half of the night. They were to measure the angles between it and its neighbours to map that section of the sky.

Each star would be observed over and over again to check the accuracy, not just tonight but on different nights by different observers and using different instruments. Then all the results would be used to calculate a definitive position for it on the celestial dome.

On a night like this, with no moon, it was so dark that the assistants used specially made candleholders mounted on poles and lines to see what they were doing. Longomontanus held his own cylinder of smoked glass close to the etched scale on the sextant and read out the coordinates for Kepler to scribble down.

Around them, soft voices uttered other numbers, and there was the occasional squeak of a metal joint as the contraptions were turned from one target to the next.

Tycho would intermittently appear from below to check on progress, dressed only in his everyday attire despite the cold. He squeezed his bulk around the overcrowded rooftop, breathing out wine fumes and occasionally supplanting an assistant to bellow out a reading himself. After the third such round, Kepler realised this was Tycho's way of helping.

When one of the quadrants over near the castellations jammed, Tycho yanked on it, creating a squeal of metal that set everyone's teeth on edge. 'We will grease it in the morning. Proceed as best you can for now,' he said, disappearing back into the castle below.

'You were with Tycho on Hveen Island, were you not?' asked Kepler.

'Yes.' Longomontanus repositioned the sextant and drew his bead along the instrument at the next target star.

'Tell me, was it always like this?'

'Like what?'

'Chaos.'

Longomontanus smiled in the dark. 'The Master has mellowed with age. On Hveen we lived with an elk for company. It was allowed to roam the corridors and feed from our tables. On cloudy nights, the drinking would go on until dawn, and the elk would drink with us.'

'Yet still you managed to work?'

'And how we worked. Wait until you see the ledger room: pages and pages of raw measurements – a vast archive – most of it just waiting to be converted into useable coordinates. I guess that is what the Master wants you for.'

'The planets too?'

'The Master has data for ten oppositions of Mars, stretching back over twenty years.'

Kepler's breath quickened. 'With those riches it would surely be possible to compute the orbit of Mars within . . . within eight days!'

Longomontanus chuckled. 'You make me feel like the wise old man, though I can scarcely be more than five years ahead of you. I have worked with the data for a long time. It is not as easy as you assume. I can reproduce the latitudes at opposition but not the longitudes.'

'Let me help you. Together, let us dethrone Mars from its position as confounder of astronomers. Let us bring it to heel and claim the gold of Egypt! Once we have Mars in yoke, so our method will harness the other planets too.'

Longomontanus eyed Kepler, his face sceptical.

'It's true,' continued Kepler. 'Mars shows the greatest differences in her speed. Determine why these differences occur and the other planets will tumble at our feet. Do you use an equant in your calculations?'

'Of course, and a deferent too.'

'I have my doubts about them. They're only needed if we assume each planet moves with constant speed. But what if they move with different speeds in different parts of their orbits?'

'Now you're being fanciful. How can the planets speed up and slow down? Do the angels that move them become fatigued?'

'I don't believe that the planets are moved by celestial intelligences within the heavenly spheres; I believe they're moved by a motive power coming from the Sun. The further this force reaches into the void, the weaker it gets. So, the further a planet's distance, the slower the planet is driven to move.'

Longomontanus's mouth dropped. He checked that the others on the roof were not listening. 'No celestial intelligences? It's as well you're not in Rome. They would burn you for that – you've heard about Giordano Bruno, I take it? Where is God in your blasphemous design?'

'God's seat is the Sun, at the centre of creation. No blasphemy there.'

'You remove him from the fixed stars of Heaven and place him at the centre of things, where damnation lies. You turn the universe on its head and say there is no blasphemy. Are you mad? There is no evidence that the Earth is moving. Tycho's arrangement is the only one that makes sense.'

'You mean that Mercury and Venus orbit the Sun, while the Sun and the other planets orbit Earth.'

'Yes, it's the only system that takes into account all the observations.'

'It's not elegant enough. All I need is sight of the measurements and I can correct the Copernican system. First, I will solve Mars and prove that it orbits the Sun. Show them to me.'

Longomontanus turned back to the sextant. 'I cannot – even if I wanted to. The measurements are the jewels of this castle and kept under lock and key.'

'Astronomy is not like ironmongery where one man makes horseshoes and another gateposts. We are a brotherhood, spread across Europe, all searching out the secret of the cosmos. We should share.'

'It's beyond my authority. I value the Master's trust above all else. Now, we must get on, there's a lot to do.'

There was a pause between them.

'Very well, but if you cannot help me, at least tell me what I must do to gain Tycho's trust?'

Longomontanus sighed. 'First, you must believe in the Tychonic arrangement of planets – not the Copernican one – and, second, you must work with him unswervingly for half a lifetime.'

Morning light fell through the domed skylight, striking a great brass globe that shone as though the smith and his polishing cloth had not one moment ago left the room. From the globe, the light bounced off to create golden murals around the circular chamber.

At sight of the monument, Kepler's stifled yawns vanished. He circled the globe with his mouth parted in awe. Drilled into the metal were a multitude of small holes around which were etched the figures of the zodiac and other constellations: the twins of Gemini with their backs turned in disdain to Cancer's nipping pincers; Orion standing proud in the opposite hemisphere to his nemesis, Scorpio. But it was the dots that truly caught Kepler's attention. He reached out to touch them, as if feeling the indentations would make them more real.

'One thousand stars.'

Kepler jumped. Tycho was inching into the lobby.

'Each one drilled into its precise position. The positions accumulated over decades, measured by my own instruments. Let no one tell you that I don't know what to do with my observations.'

Kepler felt his cheeks colour and he turned back to the globe. 'I had once thought to fashion my own contribution to astronomy in metal.'

'How so?'

'I convinced the Duke of Württemberg to commission a model of the universe for his court, based upon my nested arrangement of the perfect solids.'

'Ah, the central premise of your *Mysterium*.'

Kepler nodded. His epiphany had occurred in Graz, back in the lofty rooms of the Stiftschule where he taught geometry. Chalk squeaking on the board one day, Kepler drew a circle, then enclosed it with a triangle so that the midpoints of each straight line just touched the circle. Finally he drew a larger circle, its circumference touching each of the triangle's three points.

'In this arrangement,' he explained to the usual handful of students, 'the radius of the outer circle is twice that of the inner circle . . .'

That was when it hit him so clearly. It was as if this piece of knowledge had been woven into his soul since the moment of his birth, waiting to be remembered.

According to Copernicus, Saturn lay twice as far from the Sun as Jupiter – exactly the same ratio as the two circles separated by a triangle. Had God used such geometrical shapes as the invisible scaffolding to hold the planetary spheres in place? If so, the orbital distances of the other planets could be similarly derived by placing other shapes between them.

It had always struck him as curious that the planets were not uniformly spread throughout space. This could be the answer.

All summer he set about furious calculation. At the conclusion of his toil, he discovered that the best arrangement was to use three-dimensional shapes: a cube between Saturn and Jupiter, a pyramid between Jupiter and Mars, a dodecahedron between Mars and Earth, an icosahedron between Earth and Venus, and finally an octahedron between Venus and Mercury – and all of them centred on the Sun.

Plato had declared these shapes perfect because of the way they were constructed using precise geometrical rules. And Kepler thought he had found them mirrored in the stars, holding the planets apart. With his tutor's begrudging help at Tübingen, Kepler had published a book, *Mysterium Cosmographicum*, to announce his idea to the world. Then he set about constructing it in silver and that was when the problems began.

He explained to Tycho: 'Different silversmiths would make the various components so that no one would be able to steal the secret of the universe before it was assembled at Court. Each planetary frame would be hollow and contain a drink to be dispensed through taps at the edges of the model: brandy from Mercury, mead from Venus, strong vermouth from Mars, and a delicious new white wine from Jupiter. I even suggested that Saturn's cup should be filled with a bad red wine, so that we could ridicule those ignorant of the planet's bitter qualities.'

The old man slapped his meaty thigh in appreciation. 'What happened?'

This was the moment when Kepler regretted starting the story. 'It wouldn't fit together. What I thought were trivialities in the computations proved impossible for the craftsmen to interpret. That's how I knew I needed more precise measurements to refine my calculations.'

'And so you wrote to me.'

'Yes, sir.' Kepler felt transparent.

'Did it occur to you that your model did not fit because it was wrong?' Tycho's voice was almost as patrician as Kepler remembered his tutor's could become.

'Never. It is the only arrangement that makes sense of the Holy Trinity. God in the Sun, Christ in the sphere of the fixed stars and the Holy Spirit spread between the two.'

'So you still believe in the nested shapes?'

Kepler turned away. 'I confess my thinking has moved on. I now believe that the planetary distances may be understood using the laws of musical harmony . . .'

Out of the corner of his eye, Kepler saw Tycho raise a scraggy eyebrow.

'. . . The idea dates from the Greeks but no one has pursued it for a thousand years. As a planet moves through its orbit, so its sphere resonates and makes a note. Each planet makes a different note and together they form a divine harmony, musically rich and beautiful – for God could scarcely have designed things otherwise. Do you not agree?'

'You're speaking to a man who hasn't taken communion in eighteen years.'

'But you are working to reveal the glory of God.'

'I prefer his indifference. And perhaps that is the best I can hope for after what I have seen. I know that the heavens vary their appearance, the very realm that Aristotle claimed to be immutable – I've seen it change. The new star of 1572; I watched it for two years, blazing brightly at me yet as fixed in its position as any of the stars. It had to be located on that final sphere – the very furthest from Earth – yet who will back me? No one. The philosophers still talk of it as being an atmospheric phenomenon of the Earth.'

'I will back you. Together we can remake astronomy.'

'But we differ on whether the Earth or Sun lies at the centre of creation,' said Tycho, as if discussing a triviality, but Kepler could hear the suspicion.

'We can find common ground.'

The two men regarded each other; the only sound between them was Tycho's laboured breathing. Eventually he broke the silence. 'Tell me of your background, Johannes. Is your father a learned man?'

'My origins are not as noble as yours, Lord Brahe.'

Tycho's eyes flickered in a curious manner and he spoke brusquely. 'Nevertheless, I am interested.'

Kepler scrutinised the globe, shuffling so that its great bulk was between him and Tycho. 'I was born in Weil de Staadt. My mother is a herbalist. My father, well, my father was a soldier.'

Kepler's insides squirmed at that word – soldier. It meant only one thing: mercenary.

'Do they live?'

'My mother, yes. My father, no.' The sentence was only half correct. Aged seventeen, Kepler had watched from the shadow of a magnolia bush as the man swaggered down the street, swigging from a flagon, letting the townsfolk know that he was off to war. It was soon afterwards that Kepler discovered his father had once fought a Protestant uprising in Holland, despite his own Lutheran roots. The realisation was sufficiently painful even now to flush Kepler with shame.

That departure was Kepler's last memory of his father. When the months turned into years, the family gave up waiting. Yet it was only recently that Kepler had stopped snatching a second look at any grizzled face that passed, just in case there was a resemblance.

His thoughts were drawn back by the sound of Tycho taking a deep breath. 'I have to decide what to do with you.'

'Sir?'

'What task of calculation do I set you?'

Kepler turned at once. The Master was watching him carefully. 'Mars perhaps? Longomontanus has done more than any man alive, yet he is mired. What if I were to ask you to work with him on it? I think that might suit you, would I be right?'

A shiver passed through Kepler. Did Tycho know of his conversation with Longomontanus? He spoke carefully. 'I will not disappoint you. I will have its orbit in eight days.'

Tycho cocked his head. 'Will you now?'

'Eight days, sir, or I am not the greatest mathematical astronomer alive.'

'I once thought that I was the greatest mathematician, too. You know what I got for it?'

Kepler shook his head.

'This.' Tycho raised a finger and pointed at the metal nugget in his nose. 'I fought a duel because a classmate dared to claim superiority with numbers over me. I knew he was more gifted but I couldn't admit it. And that sliver of truth allowed his blade to slip through. The constant ache of this wretched thing is my reminder of the cost of misplaced pride.'

'Yes, sir.'

'We are a family here, Johannes. We work together.'

'I understand.' But as he hurried away, Kepler's thoughts were full only of the discoveries he was soon to make.

6
Rome, Papal States

By the time Bellarmine reached the steps leading to one of the giant doors of the Roman College, the building filled his entire field of view. He smiled inwardly; the Jesuits knew how to impress, and how to intimidate at the same time, if he were being honest. It was a formula that served them well. Even Pippe beside him had gone quiet.

The Jesuits had been in existence for only twenty years when Bellarmine had joined them, himself only just eighteen. In those days, Rome had been floundering, having still not recovered from its sacking by Emperor Charles V's mutinous hordes. It was an aimless city, its people undirected and the Pope's authority withered to a dry thread. In the face of this emasculation, the Jesuits had offered a new way forwards, a fearless way built on the intellectual mastery of spirituality, theology and philosophy.

Bellarmine thanked God every day for guiding him into their ranks. His father had hoped he would become a politician and restore the family's ebbing prestige, but his mother had seen the real path. She had believed in Church over State and had quietly urged him into the Jesuits. 'They need thinkers like you,' she would whisper to him, stoking his young ambition. 'Lutheranism is a plague; their beliefs are the buboes of evil. Only Catholicism can lead people to salvation.'

As Bellarmine had matured and studied, so he had watched Rome become strong again. As the buildings rose once more, so the people remembered their purpose, and a new determination took hold to lead the rest of Europe back to the one true Church. He also understood how the Jesuit determination had led this charge. The Vatican owed them a mighty debt. Without them . . . Bellarmine shuddered to think how far the Protestant heresies would have spread and how many souls would have been lost to the fires of damnation.

Reaching the top of the familiar steps, he glanced over at his hesitant companion. 'Don't worry, Cardinal Pippe, Father Clavius can't take your birthday away from you again.'

Pippe pulled a sour face.

They disappeared into the shadows of the entrance hall, and the heat of the day dropped immediately. Bellarmine led the way through the stone pillars supporting the domed ceiling, down the lofty corridors that skirted the courtyard, and finally up a sweeping staircase to an office on the first floor.

He knocked and led Pippe inside to where Father Clavius was waiting behind a desk strewn with papers. The Professor of Mathematics radiated concern. His squat body comfortably filled the chair, and his snowy beard lined a square jaw. Below his black biretta, his brow was pinched into deep furrows.

'I don't know you,' said Clavius in his age-deepened voice, staring past Bellarmine.

'This is Cardinal Pippe, recently appointed to the Inquisition offices,' said Bellarmine.

Clavius cocked his head, 'Dominican?'

'Yes, sir,' Pippe answered with an unusual lack of volume.

'Thank you both for coming over to see me. Please be seated.' The visitors settled into ornate wooden chairs, carved with griffin heads. 'As this is a day for introductions, let me present Father Grienberger.'

Next to him stood a giant of a man in black Jesuit robes. He was perhaps a decade older than Pippe, with an unreadable expression that Bellarmine found both compelling and unnerving in equal measures. He remained standing.

'Father Grienberger has distinguished himself in mathematics. I dare say he will follow me into this very chair when my time comes.'

Grienberger's face betrayed nothing.

'Good. Let us proceed to the matter in hand.' Atop the various manuscripts and letters on Clavius's desk was a leather-bound book. Its cover bore the marks of repeated readings. Clavius placed a liver-spotted hand on it, as if taking an oath. 'We hear from Prague that a Lutheran astronomer called Johannes Kepler has recently become an assistant to Tycho Brahe, the Imperial Mathematician to Rudolph II.'

'Why should this concern us?' asked Bellarmine.

'Kepler is a supporter of Copernicus,' said Clavius.

Pippe snorted, his trepidation forgotten. 'Can the Lutherans reach any lower? This desire to claim the heavens for human reason is abominable. To lower the planets to the realm of human wit is to diminish God's glory. Why do we even discuss it?'

Bellarmine eyed Clavius. 'I am wondering the same thing,' he said. 'I thought the ideas of Copernicus were unworkable.'

Clavius scratched his brow. 'Father Grienberger, please explain.'

'Kepler is an original thinker. He came to our attention because of his book, the *Mysterium Cosmographicum*.' He indicated the tome beneath Clavius's fingers. 'It's the first worthwhile defence of Copernicanism to be published. It's still unworkable, but Kepler has made advances in the way he treats the movement of the planets. Shortly after I saw this book, Hans Hewart von Hohenburg, the Bavarian Chancellor, contacted me with some questions of chronology. I placed him in correspondence with Kepler, to test the Lutheran's mathematical abilities.'

Bellarmine's shoulders were growing tight. 'And . . .'

'He solved everything Hewart asked of him. He is a mathematician without equal in the Lutheran Church, maybe in the whole world. Now that he has access to Tycho Brahe's measurements, he may surprise us and provide better predictions for the planetary positions than the traditional Ptolemaic method.'

'Have you discussed this with the Praepositus Generalis?'

Clavius fidgeted. 'Not yet, we wanted your theological advice first.'

'There is more,' said Grienberger. 'Tycho has observations of other celestial phenomena that cannot be explained by the traditional ways of thinking. He observed a comet in 1577 that moved through the crystal spheres.'

'Wait! Ptolemaic method? Crystal spheres? You speak another language,' said Pippe.

'Aristotle tells us that the heavens are composed of crystal spheres. The first major one contains the Moon, the second Mercury, then Venus, the Sun, Mars, Jupiter and Saturn. The eighth is the sphere where the fixed stars are located. Beyond this is the realm of the Prime Mover, which turns the whole arrangement to give us night and day. In addition to this movement, each crystal sphere has a movement of its own, which is why the planets travel across the sky at different speeds from each other and the stars. Ptolemy provided the mathematical recipe to calculate the position of the planets from the movement of the spheres.'

'Yet you say this comet moved through the crystal spheres,' said Bellarmine.

'How? How can it pass through them?' demanded Pippe, sliding to the edge of his seat.

'We do not know,' said Grienberger impassively.

'Then, Father Grienberger, you must be mistaken. They must be atmospheric phenomena. Change is possible only beneath the Moon's sphere, where the perfect ether is corrupted by human sin and wickedness.'

'We can find no error in Brahe's work.'

'Pah,' spat Pippe.

Bellarmine looked at Clavius. 'Is Tycho within our control?'

The Professor shook his head. 'His attendance at Church has lapsed.'

'Does the Emperor not insist that his Mathematician attends Mass? What did we do to deserve Rudolph II?' Bellarmine rolled his eyes.

'There is one hope,' said Clavius, looking again at Grienberger.

'Through their correspondence, Hewart and Kepler have become friends. The Chancellor now sees himself as something of a patron; so much so that Kepler sends all his letters through Hewart's personal courier,' said Grienberger.

'We could read them while they are *en route*,' cut in Pippe. 'A Lutheran writing to Catholics is surely a matter for the Inquisition.'

Bellarmine nodded. 'It would seem a prudent move.'

Clavius straightened his posture. 'Father Bellarmine, I know modesty would forbid you from acknowledging this, but you are the Church's foremost theologian; you are also a Jesuit.' The old eyes flicked to Pippe and back. 'You may wish to clarify your thinking in this matter. If a change to Aristotelian ideas is necessary, we cannot be caught unprepared – especially if it comes from a Lutheran camp. We must know if the Scriptures contain any room for reinterpretation.'

Bellarmine met Clavius's eyes. There was something fearful in them. 'I will ponder as you ask but do not be falsely hopeful. The Scriptures are quite clear in this regard: the Earth is at rest in the centre of the Universe.'

7

Benátky, Bohemia

There was little free time in Tycho's household. Kepler was permanently exhausted and his giddy spells were on the increase, too. In addition to the nightly observing sessions, there were meetings to attend, maintenance to perform and, of course, meals to endure. Meanwhile, the Mars data lay unworked like a gemstone waiting to be cut. The need to be with the figures crowded his thoughts. Every day, every meaningless chore around the household served only to increase the craving within him. And all the time, Tycho went about with his usual bombast, convinced that the act of observation was the performance, when Kepler knew it just marked the arrival of the players. The music would only come once the observations had been fashioned into a score.

Kepler skipped breakfast one morning to set down his terms of work. It was the only way forwards. Once Tycho had agreed, then progress with the data could be made. He had completed the first draft and was just dusting it with blotting powder when Longomontanus returned from the dining hall.

'I have a favour to ask,' said Kepler, swivelling from the small desk next to his cot. 'Will you negotiate on my behalf with Tycho?'

The senior assistant looked bemused.

Kepler offered the sheet. 'I have written out my requirements but I would ask that you not show the actual document to Tycho, simply discuss its contents with him.'

Longomontanus hesitated but took the piece of paper. He scanned it, sucking air through his teeth. 'No one has ever demanded so much before.'

'Because Tycho has never needed anyone as badly before.'

'I cannot represent you. You need to find another negotiator.'

'But who else can I trust?'

Longomontanus placed the piece of paper onto Kepler's desk. 'It would be better for me if you forgot that you showed me your demands. I want no part of this negotiation.'

Kepler sighed. 'Very well, I will solve Mars and then approach Tycho myself.'

That night, Kepler found it impossible to concentrate on the observations. No matter how hard the stars called for his attention, the Mars data sang more loudly. The only blessing about being on the roof was that the cold air soothed his fever, parting the fog in his mind.

A quick glance around the sky told him that his illness was probably far from over. Jupiter was shining high in the sky, raining down its influence, clawing at his liver and throwing his humours out of balance. Sitting firmly among the stars of Aries, the alignment conspired to attack his thinking. Worst of all it could last months; Jupiter would continue to creep across the sky. Only with the rise of Leo in the spring, might he hope for some strength to return.

Camomile – he must find camomile in the kitchen, and sage too. Yes, cooling camomile and purifying sage. A poultice about his forehead should balance him enough to work. Better still, a sage balsam rubbed into his torso.

It occurred to him that once he had achieved his goal and described the motion of the planets, the next step would be to understand why they moved and then how their influence propagated across space. If he could understand that, perhaps it would be possible to build a shield to bounce the celestial forces back into space – though only the ruinous ones, of course. Perhaps roofs could be made in this way and people could wear protective hats to banish illness . . .

Longomontanus called out an observation from the other side of the sextant, startling Kepler, who asked for it to be repeated and then fought the numbing ache in his fingers to record the figures. But the act felt meaningless, like collecting raindrops instead of swimming in the ocean. Any literate person could do this work, but only he could solve the shape of the Martian orbit.

In his head, he tried juggling the planet's observations. He had been shown so few that he had memorised them easily, despite the fever. Now, he could rearrange them at will and search for their hidden meaning.

'Johannes, I'm talking to you,' Longomontanus hissed. The assistants never raised their voices on the roof. It was as if they feared alerting the stars to their vigil. 'Declination: thirty-four degrees, twenty-five arcminutes.'

'It's no good,' Kepler said, letting his arm fall to his side, all pretence of standing with the notebook at the ready forgotten. 'When did an architect ever lay the stones himself?'

Longomontanus unbent himself from the instrument and pressed his hands into the small of his back. 'You misunderstand. In Tycho's castle we are cogs, not wheels.'

'I'm sorry. I have to go and work.' He did not wait for an answer but thrust the notebook at his companion and dodged the other instruments and assistants on his way to the stairway.

A few sparse torches lit the corridor beneath. As Kepler threaded his way from one pool of shivering brightness to the next, he heard the sound of hushed voices, some way ahead. The urgency of their tone halted him, and he sank into the shadows.

It was Tengnagel and Elisabeth, Tycho's eldest daughter. They were whispering and pawing each other. Tengnagel seized her by the arm and drew her close, pressing his mouth against hers.

Caught between disgust and fascination, Kepler watched the urgent fumbling of their hands upon each other. As their kissing reached its crescendo, she pushed him away and giggled breathlessly before slipping through a doorway. Kepler knew that he should move before he was discovered. Inadvertently his feet shuffled on the gritty stone.

Tengnagel was on the verge of following Elisabeth inside when the noise caught his attention. 'Who's there?'

Kepler flattened himself against the dark wall.

Tengnagel challenged again, this time drawing a small sword.

At the sight of the blade, Kepler stepped into the orange light and locked eyes with the younger man. Wordlessly Tengnagel sheathed his sword, reached for Elisabeth's door and pulled it shut. Turning on his heels, he marched away, tossing his hair.

Once safely inside his room, Kepler wrote out the memorised observations of Mars. Individually each coordinate held a glimmer of meaning; taken together they were loaded with significance.

They were the coordinates of Mars for the planet's last ten oppositions. During an opposition – according to Copernicus – Earth caught up with Mars and lapped it like an Olympian on an inside track. When this happened, every 780 days, Mars appeared to backtrack in the sky before resuming its onward motion.

Kepler toyed with the numbers, deciding on the best way to attack them tonight. At opposition Earth and Mars were at their closest, with the Sun banished to the opposite side of the sky. So, these figures provided excellent starting points for his analysis. Yet, without the observations in between, all he really possessed were a few guard towers but no walls from which to build a citadel. Nonetheless, it was a start.

Longomontanus arrived as the sky began to blush with the dawn. His movements were laboured, more befitting a man of twice his age. 'Progress?' he asked flatly.

'Some,' said Kepler. 'I need a full set of observations, from one opposition to another.'

Longomontanus rolled his eyes. 'How many times must I say this? I can show you only what the Master has allowed. Now I must sleep.' He closed the wooden shutters over the window, encasing the apartment in gloom, and then dropped to his bed where he was soon breathing evenly.

Kepler retreated to his own bed, frustrated and annoyed. It set the pattern for the rest of the week.

8

The noises started early as the servants rose to light fires and begin the breakfast preparations. A little later, the sounds of the other assistants rousing themselves after too little sleep would come from the neighbouring rooms. What began as soft voices and footsteps on the stone would inevitably rise into the occasional shout or burst of laughter. Sooner or later, someone would drop something, and Kepler would wake up.

But on this particular day, sleep gripped him more tenaciously than usual.

'Johannes, you've overslept.' Longomontanus was rocking his shoulder.

'I cannot . . .'

'You must get up. The Master is asking for you.'

'What can he be thinking? I was up again all night working for him.'

'There are strict timetables here, you know that.'

Kepler managed to hoist himself to a sitting position. He was hot and shivery, his face swollen with phlegm. Longomontanus stepped back from the stale air that escaped the bed.

'I'm unwell,' said Kepler.

'It makes no difference.'

Struggling for breath, Kepler pushed himself to his feet and reached for his clothes. He fumbled a few buttons shut on his jacket but left the doublet on the chair. He still wore yesterday's hose, so had only to slide his breeches up his legs. He teetered to his feet.

Tycho was waiting for him downstairs, in conversation with Tengnagel. As Kepler approached, they stopped talking.

'It has been eight days, Johannes. You owe me an orbit of Mars.'

'You raise me from my sickbed to mock me?'

'It was you who made the wager.'

It took Kepler a moment to realise that Tycho was walking away. He forced his aching legs to follow. Tengnagel brought up the rear, making his presence felt only by the confident rhythm of his footsteps.

They reached Tycho's study and went inside. It was a messy place with piles of letters abandoned on the desk and burned-out candle stubs on the mantelpiece, their spent wax hanging like stalactites above the hearth. Tycho reached for a metal pitcher on a table and sloshed wine into three silver goblets. 'You look dreadful, drink something.'

It was not until Kepler raised the goblet to his lips that he realised how thirsty he was. He drank deeply, comprehending too late that this was not the watered-down stuff usually served at breakfast.

'Remember, I talked to you of trust.'

Kepler set down the goblet, nodded stiffly.

'Tengnagel tells me that you met with The Bear in Prague.'

'He accosted me.'

'Quite.'

Kepler glanced at Tengnagel, then back at Tycho. 'Sir, you must not confuse a chance encounter with sympathy.'

'You have been a supporter of his in the past,' said Tycho.

'How I regret the inane letter I once wrote. Sir, you must forgive me for my naivety.'

'I wonder if the same thinking applies to me?'

Kepler fought a surge of annoyance. 'There is no one I hold in higher esteem. If you are the king of astronomy, I would willingly be your knight, to stand and serve beside you.'

'Yet you distance yourself from the work here, preferring to isolate yourself in private studies.'

'I work on Mars, the task you set me. We both know I am the one to fashion your harvest into a feast.'

'The task of which you speak would take any single man a decade or more to complete. I need a workforce to split the load and I expect you to be part of that.'

'You need no workforce to craft this vision. You just need to allow me a key to the ledger room. I must have access to your measurements – all of them.'

'Then what? You'll solve it all in another eight days?'

Kepler looked away, drumming his fingers against his thigh.

'Did we not discuss the very observations you seek over dinner yesterday?'

'What? When you mentioned the apogee of Mars in passing, then the position of the nodes in between mouthfuls. Am I to grab these titbits in the same way your tiny fool waits for bones? You think I'll be satisfied with that? How little you know of my intellectual providence.'

Tycho's face turned crimson. 'From a man whose father was a mercenary and whose mother was a witch! You would do well to remember in whose house you are standing.'

'And you would do well to remember who *you* are talking to. I am God's witness to the heavenly structure. I am the only man in Europe to unfold his design.'

'God's witness!' Spittle flew from Tycho's mouth.

'Without me, your observations are worthless!' shouted Kepler. 'No more than paper and ink!'

'Get out of my sight,' growled Tycho.

Kepler stood his ground.

'Go, I tell you!' Tycho lurched forwards, his giant body wobbling on small feet. Kepler glowered at the Dane, but a powerful hand seized his upper arm. It was Tengnagel pulling him from the room.

That night, incapacitated by fever, Kepler dreamed that Jepp entered his room. Powerless to act, the astronomer watched from some distant shore as the sinister dwarf leafed through his papers, taking a sheet with him when he left.

The next morning Kepler awoke to the whirling clatter of a carriage in the courtyard. He lurched to the window to see who was arriving. A stout man descended from the vehicle and was ushered inside.

Thank the Lord!

It was Jessenius, doubtless sent for by Tycho to sort out this fiasco. Kepler straightened his clothing in readiness but had to wait for an hour before anyone came to get him. All this time he listened to the throb of blood in his ears and his laboured breathing.

When the thump on his door came, it made him jump. Tengnagel was waiting outside. 'This way.'

Kepler was led to the study again. Tycho's round bulk was lodged behind the desk. Jessenius sat opposite. Both rose as Kepler entered. Jessenius greeted him stiffly, avoiding his gaze. Kepler could only guess at the version of events he had been told.

By contrast, Tycho was all smiles. 'Johannes, please, let us solve these silly problems between us . . . You can go,' he said to Tengnagel, who left the room with obvious reluctance. 'Now then . . .'

Kepler saw something familiar resting on top of the papers that littered the desk, a sheet of paper in his own handwriting. His stomach fell away. It was his list of demands for employment.

Tycho picked it up and started to read. 'You want me to rent a separate house in the surrounding village for you.'

'That's correct. If you insist I stay at Benátky, then I need a separate apartment, well away from your . . . court.'

'You want me to provide meat, fish, wine, bread and beer, so that you do not need to eat with us. And also firewood.'

'You provide food and warmth for your other assistants, but, in order for me to work efficiently, I need peace and quiet. I must be detached from the chaos.'

Tycho looked at Jessenius. 'All of this I might agree to, but then we come to Herr Kepler's terms of work . . . He refuses to take part in the observations and states that he will only conduct research in areas of mutual benefit. He should be free to set his own timetable, rising late if he has worked far into the previous night . . .'

'I need no spur to make my work, rather a brake to restrain me,' protested Kepler.

'. . . and for this magnanimous service he demands that I obtain a salary from the Emperor of fifty thaler a quarter.'

'There is nothing extravagant about my claims. I have heard, sir, that your great brass globe cost five thousand thaler. Such a sum would keep my family and me for the rest of our lives. I think my terms are modest in comparison.'

'If it were up to me, these terms would be met and you'd be the happiest man alive. But patience is required. I cannot simply expect the Emperor to agree to this. You must show willing.'

'How can I do that when my concentration is repeatedly broken by your unruly house? The understanding of nature cannot be shaped in the middle of a mob. Once the observations are taken, the interpretation is a solitary affair. My mind must be still, my thoughts free from all distraction, filled only with the numbers that represent nature. Then, with nothing but those numbers in my mind's eye I will see the grain of truth, the gleaming gem of reality that God has placed in us all but few know how to access.'

Something in Tycho's face changed. 'You talk of mutually beneficial research yet in reality you wish to purloin my observations for your own gain.'

'That's not true.'

'Really? Tell me, Herr Kepler, do you believe in my system of the planets?'

'I do not. You have the planets going around the Sun, but the Sun continuing to go around the Earth. It is not an elegant solution.'

'Have you no respect? I have dedicated my life to the collection of these measurements. Do you deny me the right to interpret them?'

'Just because you have taken the measurements, does not mean that you can choose which conclusions to draw.'

'But you yourself choose to ignore the observations that run against you.'

'You refer to the parallax?'

Tycho gave a curt nod.

'It is my belief that the stars are further away than we first imagined.' Kepler knew his next words would wound Tycho and uttered them with pleasure. 'Not even you can build a sufficiently precise quadrant to see the parallax. God has beaten you.'

Tycho stood up.

Kepler rose too and set his head swimming with the sudden movement.

'Gentlemen . . .' began Jessenius, also rising.

'And you have a problem,' continued Kepler. 'Longomontanus calculated the distance to Mars incorrectly. I've found the error, but it is not in the calculation, it is the observation that has been taken incorrectly.'

'There are no errors in my observations.' Tycho said. 'My assistants are too highly skilled for that. It is your mathematics that lets you down.'

Kepler felt himself begin to sway. 'I would offer to show you, though I fear your own inadequacy with numbers would prevent you from seeing the truth. The Bear was right. You will squander your life's work rather than have someone else transform it. All because you want to control that final transformation, like you control your menagerie of freaks.' His legs suddenly felt weak. The room smeared into a wash of pastel shades and his last conscious thought was of crashing to the ground.

When he came round he found himself propped in a chair near the front door. Around him, the craftsmen continued their renovations. How he got here, or how long he had been here, he did not know. He must have looked like some weird statue awaiting refurbishment, or more likely some piece of rubbish ready to be taken away.

'You're not dead then.' Tycho peered at him with those unfathomable hazel eyes.

'Forgive me, sir.'

Tycho grumbled something and waddled away. He spoke quietly to Jessenius and disappeared into the bowels of the castle.

'Come along,' said the anatomist, arriving at Kepler's side. 'Let us get you to Prague. Your things are already packed. Can you stand?'

With Jessenius's help, Kepler limped to the waiting carriage and slumped, exhausted, on the small bench seat. Again, he felt his mind depart from his surroundings, yet this time it never fully arrived at unconsciousness. Instead, it wheeled in black circles, seemingly immense turns of thought, yet never really straying far from what had just happened in Benátky.

What would he do now? How would he feed his family?

In the fog of his thinking, one name shone out.

His only hope was Mästlin, his old tutor. He must surely have received Kepler's letter by now.

9

Tübingen, Swabia

Kepler! Mästlin's stomach lurched as he was handed the letter and glimpsed the handwriting. Yet even before the flash of annoyance could subside, another emotion arose in its place: curiosity. It was six years since Kepler had left the university here, and in all that time he had written often but never returned. By all tallies that should have been long enough for Mästlin to rid himself of the troublesome man.

He should have simply dropped each letter, unopened, onto the kitchen fire and let its heat make broth, or crisp the skin on a hog. That would have been useful. Yet without fail he found himself teasing open the wax seal, unable to resist learning the next instalment.

Perhaps that was what annoyed him so much about Kepler; no matter where the young man was in Europe, he could still force Mästlin to ebb and flow, as if some tidal force linked their two minds.

This latest letter was another plea for help, and he read it with the usual mix of impatience and admiration. Pathetic and brilliant in equal measure, it was the long-dreaded application for a position at Tübingen and a request for the price of wine, bread and meat in the city, as his wife would not be content to live on pulses.

Sensing the tidewaters rising, Mästlin willed himself to head for the kitchen fire but found himself in the corridor. Even as he walked, he knew what the response to Kepler's application would be. It was madness even to pass on the contents of the letter.

The University's Chancellor was not in his office, but the smell of a recently snuffed candle hung in the air. Mästlin checked the neighbouring rooms. Those too were mostly deserted now that the afternoon was sliding into evening. Next, he stopped at the chapel. The boys were praying, enveloped in balloons of condensing breath that allowed the minister to check for miming. It brought back a memory of Kepler, hunched in devotion, always the last to release himself from prayer.

Mästlin entered the gardens, where the grass was heavy with dew and just beginning to sparkle as the chill evening turned it to frost.

The Chancellor was walking barefoot across the grass, eyes closed. Wearing nothing heavier than his usual gown, he was a spectre drifting aimlessly across the mortal world.

'Sir, you will catch your death.' Mästlin trotted over.

'On the contrary, Magister Mästlin, I find the sensation makes me feel alive. Quite something at my age.'

Mästlin regarded the shrunken man. Despite the white hair, the Chancellor was as permanent as the university's foundation stone and it was easy to forget that he was now in his eighties. Mästlin opened his mouth to broach his subject but could coax nothing at first. At last, he found the courage. 'Sir, Johannes Kepler is upon hard circumstances. He is asking whether there is a small professorship for him here?'

The Chancellor nodded curtly and resumed his walk. 'When it comes to Kepler, what astounds me most is that you continue your correspondence.'

'He may be unconventional, Chancellor, but he is also brilliant. I believe he may truly have found the secret plan that God worked to when creating the heavens.'

'By advocating the view of a Catholic?' Derision laced the Chancellor's voice.

'Copernicus only published his ideas because the Lutheran Rheticus visited him – and no doubt improved upon his work. The canon himself was too blind to see his own achievement. The new astronomy is Lutheran to the core. Kepler's *Mysterium* was the next step.'

'I seem to remember that you spent so much time seeing his book through the printers that you neglected your own studies. We can ill afford to lose your concentration again, especially with this abominable Roman calendar to fight. We will celebrate Easter when we choose to, not when the Vatican tells us. You are our sharpest voice against it.'

There was no accusation in the old voice, just a calculated disappointment that humbled Mästlin more than anger ever could. 'I will oppose the introduction of Pope Gregory's new calendar in these lands to my last breath,' he said. It was a transparent attempt to chain Europe to Catholic timekeepers and had to be resisted on those grounds, although Mästlin was secretly impressed by the quality of the Jesuit mathematicians under Rome's command.

He followed the Chancellor deep into the gardens, losing the university buildings behind the tall hedges. The lustre on the plants signalled the deepening freeze. They continued into the silhouettes of the box-cut rosemary, the old man apparently oblivious to the temperature of his bare feet crunching on the gravel path.

'The remarkable thing about age is that you learn what's important, and what can be ignored,' he told Mästlin with a faint smile.

The only light around them was coming from the sinking yellow crescent of the Moon, its illumination rendered dimmer still by its passage through the tangle of bare branches.

'I have nearly completed a new tract against the calendar, sir.' Mästlin thought of the two chapters and the scribbled notes in his bottom drawer, and was glad of the shadows to hide his face. What he needed was help from a mathematician as skilled as the Jesuits, and there was only one Lutheran who fitted that bill. 'Sir, when Kepler was here, he began theology. Entering the ministry was all he cared about . . .'

'Why did he not complete his studies?'

'You made me send him to Graz when they needed a new mathematician.'

'Did I? That was cunning of me. Ah, yes, I remember. I didn't trust him.'

'He's a grown man now. Could he at least return to finish his degree? We need men in the pulpits. The Jesuits are everywhere; founding their schools and spreading their intrigues in towns that we had long since considered our own.'

The Chancellor raised an arthritic finger. 'We need *good* men in the pulpits, Magister Mästlin. Not just any men. He was small, shuffled about a lot, used to play a woman in the university plays.'

'He is as God made him, sir.'

Something in the Chancellor's posture stiffened. 'Why must he insist on questioning our beliefs at every turn?'

'It is his nature; he means no harm. He believes, as Luther did, that to question and to reform is the foundation of our faith.'

The older man huffed, sending a balloon of expanding vapour into the night. 'Then he was born a century too late. Now is not the time for change. Our faith is minted in the currency of Wittenberg, and we cannot risk division on the whim of interpretation. Those devils in Rome will exploit any weakness.'

'On that we are agreed,' said Mästlin.

'So, the question remains: could we trust him to deliver what we tell him to preach?'

'He removed the prologue of theological justification from the *Mysterium* as requested, Chancellor. His book stands now simply as a matter of geometrical calculation, free of religious connotation and with no assertion that the Earth actually moves around the Sun.'

'I hear that Hewart von Hohenburg's personal courier handles his letters these days. The Chancellor of Bavaria is a powerful friend to have – a powerful Catholic friend.'

The night closed in on Mästlin. He stopped, the sound of their footfall replaced by the surge of blood in his ears. 'I didn't know that,' he stammered.

The moonlight sparkled in the Chancellor's eyes. 'You're not the only one who keeps an eye on Herr Kepler. There's no place for him and his heresies here, as you well know. There may be no place for him anywhere in Lutheran lands. Quite beside his useless astronomy, his dissension over the ubiquity doctrine is enough to have him excommunicated. Cut from us, he would run to whoever opened their arms.'

'Yes, Chancellor.'

'Distance yourself from him, Magister Mästlin, and brace yourself. We are at war with the whores of Rome and the bastard hoards of Jesuits swarming across our lands. If – when – Kepler converts, you cannot afford to be associated with him.'

10
Prague, Bohemia
1601

It was one of Barbara's favourite games: to unpack another trunk of possessions and lament how good their life had been in Graz. Kepler hated and avoided it whenever possible. If it were daytime, Regina would be the target, although she soaked up her mother's stories about the paradise they had left behind. But tonight there was no one else, save Frau Bezold, the old housekeeper.

How long would that pairing last before one of them raised their voice at the other? Ten minutes? Five? Kepler found that, for once, he would actually prefer to play the nostalgia game himself.

Barbara was pulling out a stack of pewter plates and piling them on the table.

'I've missed these,' she said.

'We have new plates.'

'Yes, but these remind me of our suppers in Graz. Do you remember when Ole knocked over his wine and stood up so fast that he turned over his dinner?'

'I remember the mess.'

'And the Bimeks were there, too. I wonder if his dancing has improved?'

Kepler hated this game especially, because most of the people she talked about had once been Protestants, mostly Lutheran. One by one, they had all bowed to the pressure from the Archduke and converted, just so they could remain in Graz to live out their puny mortal existence. Inside Kepler a single word was carved across his memory of each of them: traitor. He was glad to have left them behind.

The Keplers had moved into their own home in Prague in the spring. Hoffman had sheltered them for the duration of Kepler's illness, including the darkest days in November, when everyone had been convinced that the astronomer was dying. His decline had been hastened by the letter from Mästlin. It had arrived soon after Kepler's return from Benátky Castle.

'We're saved!' Kepler had cried as he pulled open the letter in Hoffman's grand hallway, bringing Barbara and the Baron running to join the celebration.

However, at the sight of the words, Kepler dropped the letter and reached for a nearby chair. Hoffman called for some beer to revive him as Barbara dived for the fluttering sheet. She read it with tears in her eyes.

I can offer you nothing but prayers. If only you had sought the advice of men wiser and more experienced in politics than I. I am, I confess, as inexperienced in such matters as a child.

'We are finished,' Kepler had whispered, feeling the strength slip from his body. Yet, somehow, his soul had remained intact. Hoffman had sent his staff skidding along the frosty cobbles into the markets earlier and earlier each week, to buy the dwindling winter supplies of Guinea spice and ginger that Kepler's fever required. Barbara fed her husband the specially prepared soups in the hope that they would purge his body. His nightly voiding of putrescence convinced them all of the treatment's efficacy and, as he shivered uncontrollably in the wake of these violent bouts, bedroom windows wide to remove the stench, so Barbara kneeled by his bed and prayed. Eventually, coinciding with the arrival of the first lily of the valley in the nearby forests, Mercury took hold of the reins and Kepler's strength slowly recovered.

Their new house was modest, admittedly. Situated on busy Karlova Street, it consisted of just two small storeys and a pair of garret rooms, but it was made of stone. No more creaking like the house in Graz every time the wind howled. Sometimes, it could be utterly silent. So silent that in the darkest reaches of the night, only the touch of the blanket convinced Kepler that he was still alive.

In the daytime, the street was livelier, with traders and passers-by. Just outside the front door, the rushing of the Vltava was within earshot. Kepler regularly marvelled at the volume of water that passed the city, especially during the spring when it carried the inland snowmelt on its way. Then, he would lift his gaze to the Imperial Palace, high on the hill beyond the river. From this distance, the people up there looked like ants, or another river flowing around a magnificent island. *What greatness went on at court?*

Opposite their house was another building that Kepler admired. Despite it being home to the Jesuits in Prague, he appreciated the curved walls of the Church of St Clement. They were swept into a perfect ellipse, truly harmonious to the eye. Wherever you went in the city, however, it was never long before you saw one of the wide-brimmed Jesuit hats bobbing through the crowd, or heard the swish of their black robes.

Once established, the Keplers sent to Graz for their possessions, and, a fortnight later, two muddy wagons arrived, drawing a crowd of onlookers.

Now that the unpacking was almost complete, the house felt as if it had been invaded. Kepler drifted within this curious world of familiar, yet unnecessary, possessions, unable to shake the feeling that his former life had caught up with him. What he had planned to be a grand new beginning had returned to the smallness of before.

Kepler looked up from his desk to see the hunched figure of Frau Bezold entering his study.

'Herr Ulmer and his son are here,' said the housekeeper.

Kepler's shoulders sagged. He shuffled to the front room carrying parchment, a quill and an inkwell.

Ulmer stood as he entered. The visitor wore an ostentatious collar that reached halfway down his chest, covering some of the food stains on his jerkin. His son raised his head only far enough to look at those around him through his eyelashes, and he fidgeted with embarrassment.

Barbara was fussing over them. 'What a handsome young man he is,' she said to the father, igniting the boy's face and setting his cheeks into involuntary movement.

Ulmer ignored her when Kepler appeared. 'What I need to know, Herr Stargazer, is the direction of my son's life; his destiny, so to speak. What manner of fortune and nobility lies in his path?'

Kepler's heart sank. He needed no star charts to tell this future. He wondered how on Earth the father could think such a timid boy was destined for any greatness at all.

'Horoscopic prognostications are not guarantees.' Kepler pretended not to see the warning look on Barbara's face. 'All I can tell is when there will be favourable celestial aspects; what the individual does during those times is up to him.'

'There are others in the city who claim to be able to predict exact events, down to the very day. Cheaper, too.'

'Fraudsters! They can do none of those things. No one can.'

'Then why am I here?' Ulmer lifted his whiskered chin.

Kepler pulled his own goatee into a point, letting the wiry hairs spring away from his fingers. 'Sit down, please, Herr Ulmer.'

From across the table, Kepler explained, 'Inside us all is an imprint of the heavens at the moment of our birth. When this is matched by a similar aspect in the heavens, so our souls resonate and our true natures are brought to the fore.'

'But is that good or bad?'

Barbara was watching her husband almost as intently as Ulmer as he set down his paper and writing tools.

'It depends upon the aspects at the moment of your birth. If the natal arrangement was less than propitious, you are better to wait until the alignment tempers your natural inclinations.'

'How are we to know?'

'That is what I can help you with.'
Ulmer nodded emphatically. 'Then let us begin.'
'I'll fetch wine,' said Barbara, calling for Frau Bezold.
Kepler reached for his quill, flipped open the ink well and looked at the boy. 'Date of birth, please?'
'The eighth of August, 1582,' the father answered.

Ulmer and his hapless son departed some time later carrying Kepler's assessment. As the astronomer had thought, there was little opportunity for greatness but in three years' time there would be a small window of opportunity, when Jupiter would fall into conjunction with Mars.

Kepler hoped that the red planet would muster some energy in the boy and Jupiter's influence would steer him to a modicum of leadership, probably through marriage and taking control of an estate. Kepler had talked up the possibility, and the father had bounced out of the house upon the news.

As soon as the door was shut, Barbara threw her arms around Kepler. 'Two whole gulden for a few hours' work!'

Kepler silently wished he could share her enthusiasm.

'I have more good news for us,' she said, dragging her words as if unsure about how to phrase her revelation. 'We are to be parents again.'

'Truly?'

She nodded, her cheeks rosy.

'We must praise God.' He clasped her tightly as though he would never let her go.

The rancid smell of burning tallow filled the little study. Kepler had grown accustomed to the odour, insisting that the expensive beeswax candles were kept solely in the front room, and then only lit when they had supper guests. Now, the unpleasant tang of tallow was integral to setting his mood for work.

He looked once more at Copernicus's calculations and pushed them around the page. He added them, subtracted them, multiplied and divided them in his attempt to wring out some more meaning from the chimeric figures. But it was no use, the measurements had been forced together from such disparate sources that Copernicus himself had doubted their veracity and even dropped the figures that did not serve his purposes.

Kepler reinstated those outliers, unwilling to doubt them on someone else's say-so. Yet, even putting them back in, he made no headway. Each number was the eye of a needle; the correct orbit would thread every single one. Copernicus had threaded some, missed most. Even Ptolemy and his ugly Earth-centred universe could do better. But how could that be right – how could the entire vault of Heaven turn once a day while the puny Earth remained stock still? It was absurd; it put everything backwards. So why could he not prove it? What was wrong with him? Kepler found himself watching the flecks of soot as they danced in eddies above the candle and thinking of Tycho.

The door shot open, rocking on its hinges. 'We must talk. The housekeeper is being insolent again.'

'I'm busy, Barbara.'

'She claims not to have enough money to buy food.'

'Is she right?'

'We have the Dietrichs coming on Friday – we cannot give them bread and sausage.'

'Quite.'

'I want a sheep's head for the centrepiece, decorated with the entrails.'

'Is it so important?'

'It's what people of our standing eat these days. Husband, your head has somehow passed clean through our station and lodged itself firmly in the stars.'

'Then we'll have to borrow against next month. The merchants know we're good for credit.'

'We've already used up all our credit.'

Kepler ran a hand over his face. 'Then we'll have to sell some possessions – quietly, so no one knows. What about that small table you keep your prayer book on. Do we really need it? You could keep the book on the mantelpiece.'

Barbara's nostrils flared. 'Don't you dare touch that table.'

'Then something from upstairs, something that won't be missed when we have visitors.'

'Understand this, Johannes Kepler. We're not selling one single item of mine from this house. It's all I have left after you forced me to leave behind my family and friends. Why not sell something that you brought into this marriage?'

'But I had nothing,' he stammered.

'Precisely. This is a problem of your own making. It's up to you to solve it.'

'But how?'

Barbara bunched her fists and planted them on her hips. 'You must take more clients.'

'Oh, Barbara, you know I hate it. Astrology is not some conjuror's trick. It should be put to a noble purpose, not telling fatheads what they may or may not do with their little lives.'

'Noble purpose? What do you know of noble purpose? No one can understand your stupid book with all its shapes and signs. And it's not put any food on our table.'

'Five planets. Five perfect solids. They have to be linked. It's obvio— Oh, what's the point? You can never understand my work. It's beyond your grasp.'

Barbara blinked at his words. He thought for one awful moment that she was going to cry. 'Maybe not,' she said, 'but I *can* understand poverty and hunger.'

She rushed from the room, the slam of the door reverberating through the house.

Next morning, Barbara picked at her food. Her feeble sips of wine made Kepler feel uncomfortable, and Regina looked from him to her mother, divining some tension but unable to comprehend it.

When Barbara did take a mouthful, she gagged on it and went running for the back door. She heralded her return with a complaint to Frau Bezold about the quality of the meat.

Kepler moved round the table and held out her chair. 'Come and sit down.'

From outside, a commotion drew their attention. The clop of hooves and the trundling of a multitude of carts grew in volume. Wagon after wagon passed the window, all packed high with cloth-covered burdens. Outriders on horseback trotted by.

'It's a festival,' said Regina, racing to the front door.

Her parents followed her into the street to watch the procession.

But there were no acrobats, no exotic animals, no men on stilts, no women dressed as goddesses; just donkeys and carts laden with people and possessions.

'Someone important is moving into town,' said Barbara, craning to see into the carriages.

Kepler did not hear her words. His full attention was focused across the street to where a dwarf in a jester's outfit was pulling faces at the onlookers, occasionally jumping at them, as if he intended to attack.

'Take Regina inside. I must find Jessenius at once.'

Barbara hesitated.

'Do it!' Kepler began running, passing the curve of the Jesuit College and outpacing the lumbering wagons. He headed along the banks of the Vltava, where people were stopping to look at the procession. Pausing for breath, he looked back with them. The caravan was making its way across the Stone Bridge, heading for the Imperial Palace. Kepler pushed through the crowd and set off again.

He slowed only as he neared the university. Panting, he made his way inside and, after a brief search, found Jessenius in a room of preserving jars. The anatomist was lifting one to the light so that he could study its ghoulish contents.

'Tycho is in Prague. It looks as though he is here to stay.'

Jessenius turned from his inspection. 'Have you not heard? You do bury yourself away sometimes.'

'Heard what?' asked Kepler.

'Ursus is dead. Tycho is the new imperial mathematician. He is moving into the Golden Griffin, beneath the Palace, on Hradčany Hill.'

'Jan, I beg you. Arrange a meeting with him for me. Let me persuade him to give me a second chance. I have spent this year selling fortunes. My brain withers without the challenge of his observations.'

Jessenius set down the jar, the unidentified clump inside jostling from side to side. 'But you know how disastrous it was last time.'

'I was ill, my mind was not my own.'

'On the day I brought you home, Tycho whispered to me that all could be forgotten with a letter of apology yet you sent him one of abuse.'

'I was raving with sickness, you know that. I behaved like a dog. But I am cured now, no traces of fever for months.' Kepler took a step forwards. 'Please, Jan. No preconditions this time, I will do anything. Barbara is with child. I need a sponsor.'

Jessenius ran a finger across the tabletop. 'Very well. I will see what I can do.'

11

Kepler thought it fitting that Tycho's household should take up residence in a former inn. Indeed, the Golden Griffin's serving room was the first location he saw when Jessenius held open the front door for him that morning. The wide space was lined with faded wooden panels and its trestle tables were still strewn with the remains of breakfast. An overturned goblet had spilled its contents onto the table, and the stain offered a hint of the wood's original colour.

Two servants were clearing away, piling up the leftovers and stacking the plates. One girl was tiny, easily mistaken for a child; the other was buxom, with wide hips. The men shook the rain from their hooded cloaks, drawing their attention. The big one ambled over and took their wet garments.

'We're here to see Tycho,' said Jessenius.

'Who isn't?' she muttered and listlessly went to announce their arrival.

'Remember, Johannes, I'm taking a risk here for you today. One hint of your temper . . .'

'I know.'

'Tycho is now the most powerful mathematician in the world. You cannot upset him.'

'I know.'

'It's just that . . .'

'Jan, I know.' Kepler turned. 'I've learned my lesson.'

As Tycho approached them Kepler noticed that his face was fleshier than before and his overall bulk had increased. At his neck he wore a set of heavy gold chains that seemed to pull him into a stoop. A stiff-legged stalk had replaced his waddling gait. In his rounded breeches, his legs resembled giant upturned hams as they tapered to unexpectedly delicate ankles.

Tengnagel was just behind him; there was no spare room abreast in the corridor, but as they entered the bar, he took hold of Tycho's elbow. 'Allow me, sir.'

'I don't need your help.'

Tengnagel dropped back, contenting himself with a glare at Kepler.

Tycho stopped a few feet in front of them and greeted Jessenius. Then he faced Kepler and stepped forwards.

It was still impossible for Kepler to read those hazel eyes but it was no effort to hold their gaze. After a moment's appraisal, Tycho reached out his arms and drew a surprised Kepler into a bear hug. The Master smelled like food on the turn.

'Welcome, Herr Kepler. I have missed you.'

'And I you.' A wave of emotion enveloped Kepler. He masked it with some deep breaths.

They released each other and stepped back.

'I see that your appetite has not improved this past twelve months.'

'Astronomy and the love of God are the only food I crave, sir.'

Tycho laughed. 'Do you know, I believe you.'

Jessenius departed, looking relieved. When Tycho asked Tengnagel to leave them alone too, the young man drilled his gaze once more into Kepler before strutting away.

'I have something to show you, Johannes.' Tycho led the way to an antechamber inside which were a number of statues, their details masked by white sheets that covered them and reached to the floor.

'These used to line the walls of my study at Hveen,' said Tycho. He snatched at the first sheet and jerked it upwards. It was a marble statue of an ancient Greek scholar. The inscription read TIMOCHARIS.

'The Alexandrian who compiled mankind's first star catalogue,' said Kepler.

Tycho revealed another.

'Hipparchus,' said Kepler. 'His star catalogue showed that there was a precession of the equinoxes.'

Tycho nodded. He moved to the end of the line and unveiled the penultimate statue, watching Kepler carefully.

Kepler found it difficult to disguise his surprise. 'It's you.'

'Indeed it is. Am I not the greatest star charter in history? Do I not deserve to stand alongside these others immortals?'

'Then who is the final one?' asked Kepler, with a fluttering sensation in his stomach.

Tycho seemed to hesitate before revealing the last figure. It took Kepler a moment to understand what he was seeing and to realise with shame what he had vainly hoped it would be. It was not a statue of him, but a caricature of Tycho: broader-shouldered with strong features, a proud jaw and a piercing stare.

'Tychonides, my once imagined son. The person I dreamed would mould my observations into a new system of the planets. Ever since the duel, I knew that the person to make the best use of my observations would not be me. I hoped for a son and have been blessed with three, but not one is a mathematician; the curse on the Brahe line continues.'

Kepler walked around the statue as he listened to Tycho, admiring the quality of the work. Something in its sculpted gaze resonated in his soul.

'So, I have to look to my assistants. There is Tengnagel, of course, but he lacks the application – he's to become my son-in-law, you know, God help us all. There is Longomontanus, though I fear I will soon lose him to Copenhagen. So, really, I am still searching for my Tychonides.'

'Sir, I cannot match you in birth or breeding, but I can be your humble servant,' said Kepler.

A sound at the door drew both men's attention. Tengnagel stood there, undisguised curiosity on his face.

'Sir, we need you to direct us about which room is to be your study . . .'

'Not now, Tengnagel. Didn't I tell you I was not to be disturbed?'

'It's important.'

'Just wait!' Tengnagel dropped his gaze but remained at the door.

Tycho looked back at Kepler. 'Are you a changed man?'

'As night turns to day.'

Tycho fiddled with the limp ends of his moustache and then released them with a click of his tongue. 'We must present you to the Emperor. Get you a worthy salary.'

Tengnagel's mouth hardened.

The wind in Prague followed the same course as the river, and Kepler was in its way. It sliced through him as he crossed the Stone Bridge, head down, watching his feet on the flagstones.

Above the constant hiss of the air, he thought he heard something. It came again, unmistakable this time. It was the rhythm of feet marching in time. Kepler whirled around.

Soldiers strode towards him, less than a dozen paces away and gaining fast. They carried halberds at the ready. People jumped from their path.

'Make way for the Emperor's business!' the leader shouted.

Kepler leaped to the side of the bridge, pressing himself against the cold stone, but they did not give him a second look. Behind them, four burly men trotted carrying a large pole on their shoulders, two runners at either end. They were sweating, despite the cold. Suspended from the pole was a large packing case. Its wooden surface betrayed the marks of a long journey. A second set of soldiers clanked past, bringing up the rear. Kepler watched them dissolve into the New Town, heading for the Palace.

Arriving at the Golden Griffin, he heard Tycho calling from a carriage.

'Come on, come on. We can't be late.'

'I'm perfectly on time.' Kepler climbed in.

'Maybe for a social visit. But not for the Emperor.'

Tycho swung the door shut and thumped the carriage roof. It jerked into motion.

The carriage lurched and bumped its way up the hill, finding every pothole in the road. Each fresh jolt ricocheted through Kepler's bones. One bump was so violent that Kepler thought he heard the wooden floor splitting.

Tycho groaned, as if someone had just kicked him in the backside. 'This is the final indignity of old age, you know. You have to ride because you cannot walk yet the ride itself is intolerable. Why can't someone invent a coach that doesn't try to shake what remains of your life out of you?'

The carriage – and Tycho – grumbled on until they stopped at the Palace's western gate. A soldier peered in at them and waved them on. Their journey ended in an inner courtyard, where a smartly dressed man was hurrying over to meet them.

'You're late. I was just beginning to worry,' he said, sweeping his fringe away from his face.

Kepler avoided Tycho's look as they stepped out of the carriage.

'Mister Brahe, it is a pleasure to see you again.'

'How did His Majesty receive my last horoscope?'

'With pleasure and interest, as always. He thanks you for your foresight. You must be Johannes Kepler. I am Johannes Wackher von Wackenfels, one of the imperial councillors, at your service,' he said with a bow, opening his arms as he dipped his body. 'Would it interest you to know that we are distantly related, through your father's side?'

Kepler searched beyond the long fair hair and the pale skin for any hint of family resemblance.

'There is noble blood in you after all,' said Tycho, hijacking Kepler's attention.

'I assure you, gentlemen, whatever nobility there once was, has now sadly departed in all material ways. What remains is etched in my spirit,' said Kepler.

During the negotiation for Barbara's hand, his mediators had caught a whiff of the knights in his lineage and used it to persuade her father. When the old man had discovered it to be only half true he broke the agreement. Only intervention by the Lutheran leaders in Graz, who had originally approved the match, had shamed him into reinstating the marriage.

Kepler turned to von Wackenfels's expectant face. 'That said, I am doubly pleased to make your acquaintance, cousin.'

Von Wackenfels looked as though he would burst with joy. 'This way, gentlemen, there's not much time.'

Tycho stumbled along as fast as he was able, giving Kepler plenty of time to gawp at the tapestries full of hunting scenes or biblical stories. Hard as it was for Kepler to believe, the capacious chambers and walkways made Tycho's Benátky look small.

Through the warped glass of the windows, Kepler spied more buildings. Some were large enough to qualify as whole castles in their own right. They stretched back and back, all within the confines of the Palace walls.

'This is a town in its own right.'

'Welcome to our humble abode,' said von Wackenfels.

They emerged into a bustling hall, as big as a cathedral. Kepler gasped at the sight. The great gothic vaults reached up perhaps sixty feet into the air. Each archway tapered to a point from which ribbons of plaster curved away to run in sensuous lines back to the floor.

Great chandeliers, each one a fantasy of metal and wax, hung tethered from the ceiling. Kepler estimated a hundred candles in one alone before his attention was drawn away.

Light spilled in from the giant windows, bringing the bright hope of morning, and the place was filled with courtiers. They gathered in pools of conversation or wandered in pairs through the magnificence, filling the place with lively chatter.

Until today Kepler had never seen such exquisite clothes, surpassing even Hewart von Hohenburg's garments, or such beautiful people. Each one of them was encased in a vibrant colour, tall and upright, with bright eyes and shining tresses.

'Do not wear black,' Tycho had advised. 'That's the Emperor's colour.'

Kepler was wearing navy blue, freshly, if haphazardly, laundered by Frau Bezold. Upon dressing he had delighted in the luxuriant feel of it and felt self-conscious as he stepped into the street. Now, he realised, he barely made the grade. It was not that the cloth was inferior, but the level of adornment was lacking. He felt drab in comparison and regretted arguing with Barbara over her suggestion that he buy a new suit for the occasion.

At the very end of the hall stood a pyramid of steps, its top truncated to provide a wide dais. On it, sat a single empty throne.

'If you would like to wait here, I will make sure His Majesty knows that you have arrived.' Von Wackenfels bowed and left.

Along the sides of the hall, booths had been constructed where paintings and other *objets d'art* were offered for sale. Courtiers milled around, scanning the works, commenting on them and occasionally parting with money. Kepler sauntered over to take a look and was confronted by a number of shocking depictions. Mingled in with the traditional landscapes and city views – as if all was perfectly normal – there were portraits of such lewdness that Kepler had to walk away.

Tycho laughed at him. 'The Emperor is a man of some passion.'

They waited. Occasionally someone would talk to them, pass the time of day and then move on. Mostly they waited. Time stretched into first one hour and then another.

Tycho stood stiffly, his face a mask. He leaned over. 'I have to piss.'

'Shall I find von Wackenfels?'

'I might be old but I'm still capable of pissing on my own. It's all I seem to do these days.'

At that moment, von Wackenfels approached. 'His Imperial Majesty will see you now.'

'Is everything alright?' asked Kepler.

Tycho cut across him. 'Most kind of His Grace to see us so promptly.'

As they walked towards the throne, von Wackenfels sidled up to Kepler. 'His Majesty had a new painting delivered this morning. He wanted to spend some time with it.'

Rudolph appeared and the room hushed. People positioned themselves and bowed. Kepler followed suit, bowing as low as he could. From the corner of his eye, he saw Tycho struggling to curve his body, and knew better than to offer assistance. There were beads of sweat on Tycho's brow as he straightened.

'Your Majesty, Tycho Brahe and Johannes Kepler,' said von Wackenfels after Rudolph had seated himself.

Kepler followed Tycho to the base of the steps. They bowed again.

Rudolph's face was all jowls and chin, across which grew a luxuriant auburn beard. At its centre protruded two fulsome lips. His dark eyes looked glassy and remote, and he gave no acknowledgement of the men. Kepler was uncertain whether he was even looking in their direction.

Tycho spoke. 'Your Majesty, I intend to publish my life's work as a set of tables with which any astronomer may look up the positions of the stars and the planets. It will become the world's standard reference for astronomy and will form the basis of all future almanacs. You would bestow upon me an inestimable honour if you were to permit me to call this work *The Rudolphine Tables*.'

Rudolph's face did not move for some time, and Kepler sensed Tycho's uncertainty. When it did move, the voice was a soft mumble: 'It is acceptable.'

'Thank you, Your Majesty. With your name attached, it cannot help but ensure that the work will be remembered throughout all future time. As I do not need to tell Your Majesty, this undertaking is a difficult task. I have a lifetime of observations to work through. I would humbly beg that, to assist me, Your Grace employs this man, Johannes Kepler.'

Rudolph mumbled again. 'Can one man make such a difference?'

'Only if it is this man, Your Majesty. There is no one like him in the world. His gift with numbers is incalculable.'

A childish squeak escaped Rudolph. It was followed by a stifled laugh that shook his body. 'His gift with numbers is incalculable. That's very good. You're a dry wit, Mister Brahe.'

'Thank you, Your Majesty.'

'Johannes Kepler, how do you find Prague?'

'It is the beating heart of Europe, Your Majesty. Your great works here influence everything.'

'Indeed they do.' The Emperor returned to somnolence. At last he said, 'It is agreed. You will be paid from the imperial purse.'

The ledger room was everything that Kepler had imagined: shelf after shelf of leather-bound observations. For the stars, the collection was categorised by season and subdivided by declination, the latitude on the sky. For the planets, a separate shelf was devoted to each one of the five orbs.

There was just enough space in between the cabinets for a table and chair, and, even though he was assigned a separate study, it was here that Kepler preferred to work. Surrounded by Tycho's treasure, he spent hours leafing through each ledger, acquainting himself with the data, growing to know the pages until some were as familiar as old friends. Within a week he could reach to the correct corner of the room for any observation he required. A fortnight later, he could point to the exact ledger.

The only actions forbidden were to make copies of the observations, or to remove them from the room.

'You finally achieved everything you yearned for.' The tall figure of Longomontanus stood in the doorway.

'I am helping the Master to compile his tables, nothing more.'

Longomontanus raised an eyebrow. 'You will work on Mars, I know you well enough by now.'

Kepler slowly closed the ledger. 'What of you? They say you are leaving.'

'I am away to Copenhagen. A professorship awaits.'

'A professorship?' Kepler's thoughts turned fleetingly to Tübingen.

'I have been away from my homelands too long.'

Tübingen appeared again in Kepler's mind. 'Then I wish you good fortune, my friend. We do part as friends, do we not?'

Longomontanus brought his hands together as if he were about to pray. 'Johannes, the Master's arrangement of the planets may not be complete but it is the best we have. It fits the data. Why tear it down and start again?'

'Because, in my heart, I know it is wrong.'

'Then I must resist you every step of the way. For, in my heart, I know it can only be right.'

Later, another visitor passed the room: Tycho. He rested himself against the doorframe. 'Still here?'

'Working late,' said Kepler, rubbing his eyes.

A smile tugged at Tycho's pale lips. 'Of course.'

Soon afterwards, Kepler rose from his seat and stacked his papers. He carefully locked up the ledger room and turned for the front door. It was a chill night, so he headed back inside towards his study, where he had left his hat, gloves and cloak.

The household was asleep. The long nights of observation were at an end now that Benátky had been abandoned. The instruments were in storage and the catalogues were as complete as they would be.

Kepler crept through the stillness, entered his study and froze. A small shadow was at his desk, rifling his papers. It carried a single small candle.

'Leave my things alone,' said Kepler evenly, recovering from the initial fright.

Jepp turned slowly. There was a control about his movements that Kepler had never seen before.

'You're not the imbecile you pretend to be, and you're certainly no seer. The only foolish thing you have done in all the time I have known you is to think that you could fool me. Now get out.'

Jepp took a step forwards. There was a cold clarity in his eyes that took Kepler by surprise. It was strangely entrancing. A dart of movement caught Kepler off guard. Before he knew what was happening, the candle Jepp had been holding was flying through the air.

He batted the flaming object to one side, sending hot wax cascading around the room. The candle bounced from the wall to the floor, narrowly avoiding a stack of paper. He pounced at once to extinguish the flame.

Shaking in the sudden darkness, Kepler looked around.

Jepp was gone.

12
Rome, Papal States

Claudius Acquaviva went by several names. Hardly anyone used his given name because of the station he held. He was head of the Roman College, leader of the Jesuits and one of the chief advisors to the Pope. Those who thought of the Jesuits as soldiers of God fighting a war of ideology against the hated enemy referred to him as Praepositus Generalis. Then there were those who feared him.

There were many who fell into this category both across the Lutheran world and even within the Roman one. They worried that being Catholic was one thing, being Jesuit was altogether another. They considered him scheming and possessed of dark motives. To them he was covertly known as the Black Pope.

He was a skeleton in black shrouds. Though not yet past his fifty-fifth year, Claudius Acquaviva's head was no more than a skull with a tight covering of skin. His hair and beard were shaved to stubble, and the sharp outline of his cheekbones were clearly visible on either side of his angular nose.

He worked in the shadow of a crucifix, a six-foot-tall wooden carving of Christ's death secured to the wall behind his desk. In the far corner, behind any guests, was a hooded falcon, kneading its perch. He would send the bird of prey soaring over the rooftops from the window when he needed to clear his mind.

Acquaviva's eyes, as black as his clothing, floated in brilliant whites that even Bellarmine found difficult to meet for any length of time. They were looking at him now, having lifted from a sheet of writing that Bellarmine had handed to him a few minutes earlier.

'This is copied exactly?' Acquaviva's voice never rose above that of softly shifting gravel.

'Exactly as it was written to Herr Kepler, Father General. Tübingen have refused to help him find a professorship. He's isolated. His most powerful friends are now Catholics. He may be worth . . . approaching.'

Acquaviva remained impassive. The Praepositus Generalis's office closed in around Bellarmine, who added quickly, 'It would not be the first time a Lutheran scholar has converted to us.'

'Indeed not,' said Acquaviva. 'But why would we want such a troublemaker? Father Clavius has informed me of Kepler's astronomical ideas. They are strictly against Aristotle, and we are sworn to obey the orthodoxy.'

'For control, Father General. By making Kepler a Jesuit, we could steer his efforts away from the heresies. He's a pious man; he'd see reason.'

'Just because he has been refused a professorship by his old masters, why should that make him ready to renounce Lutherism?'

Bellarmine felt a rush of pride. 'There is something else. We've learned that he has not signed the Formula of Concord.'

Acquaviva inclined his head. 'In truth?'

'Most assuredly, Father General. He disagrees with Lutheran dogma over their ubiquity doctrine.'

Acquaviva smiled faintly. The Formula of Concord was the latest German effort to agree a succinct statement of Lutheran beliefs and thus shore up support across Europe in the face of the Catholic resurgence – and it was failing. Sweden, Denmark and England were protesting against it. Not even in Germany could full support be found; the Lutheran communities in Hesse, Zweibrücken, Anhalt, Nürnberg and others were refusing to sign. Now a prominent thinker from Tübingen was turning away. This was indeed a wedge to be pounded into a split.

Bellarmine could see from the ghost of a smile that Acquaviva grasped the importance. It was the most emotion Bellarmine had ever seen him show.

'Can you spare the time to visit Prague?' the Praepositus Generalis asked.

'With great respect, Father General, it would be better if Father Grienberger made the journey. He is known to Kepler; they have exchanged letters.'

'Then Father Grienberger it will be. While he's there, he can remind Rudolph II of his duty to Rome. The reports suggest that most in Prague now prefer worshipping with the Utraquists, and that the surrounding lands are mostly Lutheran.'

'The Utraquists have signed a treaty with the Vatican.'

'Your diplomacy does you justice, Cardinal Bellarmine, but I think we both know the Utraquists are soft. Their insistence on treating the laity with the same dignity as the clergy is contemptible. I hardly need to remind you of all people that in matters of faith there can be no compromise.'

From over his left shoulder Bellarmine could hear the falcon clawing at its perch. Acquaviva slipped on a leather glove, crossed the room and unlaced the bird's tether. It automatically stepped onto his hand. He stroked the bird before unlatching the window. 'See to it that more pressure is brought to bear on Prague.' He slipped off the falcon's hood and flung the bird from his hand, out into the air. 'The sooner we re-establish direct control over the so-called Holy Roman Empire the better.'

Bellarmine watched the flash of the bird's feathers high above the city.

'It is time for Rome to spread her wings again,' said Acquaviva.

13
Prague, Bohemia

Kepler ran his finger along the shelves and drew out one of the Mars ledgers. As he did so, something quickened inside him. Longomontanus was right; he would never be able to leave Mars alone until he had cracked the orbit.

Tycho was out, summoned to court for a reading of the imperial stars, and then to supper with an imperial councillor. The rest of the household seemed content to spend their evening in the bar room with the servant girls. Kepler would have hours to indulge himself.

The book was stiff to open, as if it had not been handled in a long time. He sat down with it and scanned the figures. In his mind's eye he could trace the journey they were describing. Mars was in Leo, gliding towards conjunction with the bright star Regulus, so that both could shine their red light down on Earth.

He began copying some figures onto a rough sheet of paper, savouring the curve of each character. It was the best way he knew to become acquainted with them. By writing the numbers, he brought them closer to him. Each number was an individual. There were the ordinary and the eccentric, the important and the seemingly worthless. Each one related to the others as every man could be traced back to Adam. He would tease out the relationships. He would look for patterns. Which numbers were doubles or triples of another? Which were fractions? Soon he would glimpse the furtive look that passes between lovers at a party and betrays the secret of what is really going on.

The immensity of his task confronted him. How he regretted having once boasted about solving Mars's orbit in eight days.

How could he have been so stupid? If he was going to achieve his goal of looking down on the orbit of Mars from God's own viewpoint, to actually watch the planet following its path around the Sun then he would first have to take into account that Earth was moving as well. As Earth travelled on its own journey, so it changed the perspective from one observation to the next. To deal with this, he would have to compute the movement of Earth and subtract it from Mars. And he would have to do it for each and every one of the observations he wanted to use. Otherwise the numbers he had in front of him were meaningless.

It was a step that no one had to take in either Ptolemy's or Tycho's vision because their Earth stayed fixed in the centre of the cosmos. Only in the Copernican view of a moving Earth did the correction have to be made and it more than doubled the workload.

Once he had the true Martian motion, then he could begin searching for a curved shape that would fit through all the points. All shapes – triangles, squares, circles, everything – could be described mathematically; this was the basis of geometry. But in God's Heaven, the only reasonable shape was the perfect circle because a planet on the circular track would always stay the same distance from the centre of the universe. Yet Copernicus had not been able to fit a circle to his observations. Why not?

It had driven Copernicus's student temporarily insane. Rheticus – so the story went – became so confused that he appealed to his guardian angel, beseeching the divine creature to appear as an Oracle and reveal the answer. However, the spirit grasped Rheticus by the hair and slammed him into the walls and ceilings of his workroom, shouting, 'These are the motions of Mars.'

The last thing Kepler wanted to feel was that he was banging his own head against a stone wall. It would have to be a delicate analysis. His concentration was broken by an almighty commotion from the bar.

Jepp was squealing, 'The Master! The Master is coming!'

So early. Kepler hastily closed the book and returned it to the shelf. Something must be wrong.

Kepler left the room and crossed the short distance to the hall. People were milling around but there was no sign of Tycho. He approached someone whose face he could dimly remember from Benátky. 'Where's the Master?'

'He's not here yet, but we've learned to trust Jepp's powers.'

Kepler rolled his eyes and turned back for the ledger room, but the sound of a carriage stopped him. It grew louder and then drew to a stop. Jepp flashed past and ran out into the dark. When he returned, it was at Tycho's side, the Master's hand resting on Jepp's closely cropped hair as if favouring a pet.

Tycho was walking with more difficulty than usual, sweat running down his face. 'Fetch me a piss pot,' he bellowed. 'And hurry, or I'll choose a hat.'

A servant returned with a chamber pot. Jepp disentangled himself.

Tycho turned from the crowd and dropped the metal bowl onto a nearby chair. Grunting with exertion he loosened his belt and thrust down his breeches. He took deep breaths and placed a hand on the wall to steady himself. He took aim and waited.

Behind him, the crowd stayed rooted and silent, collectively unsure what to do for the best. As they lingered, Tengnagel arrived from upstairs, buttoning up his jacket.

Tycho took more deep breaths. Then yowled in pain as a small stream set the bowl ringing. An acrid smell filled the room. He tried several more times before giving up. He tucked himself away and buckled up roughly. 'What? Are you all just going to stand there?' he roared at their gawping faces.

The paralysis broken, the members of the household scurried here and there, looking for the quickest way to leave.

'Johannes,' barked Tycho.

'Yes, sir?'

'You have left the ledger room unlocked.'

'Yes, sir.'

Tycho lurched into motion, heading for the interior of the house. 'Where's my wife?'

The next morning the household was still at breakfast when Kepler returned to the Golden Griffin. He looked into the room hopefully but Tycho was not at the table, and there was a grey pallor over the proceedings.

He walked to Tycho's study and found him there, sitting behind his desk, lost in thought and still dressed in the same clothes as the night before. Kepler hovered. Tycho looked up and waved him over.

'How do you feel?' asked Kepler.

'Hungry, if you must know. My wife in her feminine wisdom has forbidden me to eat until I have cleared myself out. So, I sit waiting for my body to rouse itself.'

He took a small cup from the table and swirled its contents. 'It's supposed to do me good. Brandy would do me more good.' He shouted at someone unseen, then knocked back the drink with a grimace and wiped his mouth along the sleeve of his doublet. 'How soon before we have the first pages ready for the printer?' he asked, clattering the pewter cup back onto the table.

The question took Kepler by surprise. 'It is the earliest of days yet.'

'I should at least buy the paper though, do you not think?'

'If it pleases you.'

Tycho's study was too tidy, not at all as it had been at Benátky. There were too few papers and they were too neatly stacked. His most precious books were resting in a bookcase, rather than strewn across the desk. He poured himself some wine and drank it straight down. 'When did you know that you were an astronomer?' he said.

'Sir?'

'It's an easy enough question: when did you know that you were an astronomer?'

'You mean, when did I realise that I wanted to be an astronomer?'

'No.' Tycho banged his palm upon the table. 'Men like us don't *want* to be astronomers. You and me, we were born to it. The stars implant themselves at our birth, and wait to be triggered. When was it for you?'

Kepler smiled at the notion, Tycho's words sparking a memory as pure as the stars on a frosty night. 'I was six. My mother took me to a hill just outside town and showed me a comet; she taught me not to be afraid of them.'

'The bright comet of 1577?'

'The very same.'

'I saw it too, tracked it across the sky.'

'That's when I became interested in the stars. At university, I would defend Copernicus in any debate that blew up, but my original inclination was for the ministry.'

Tycho tutted. 'Who taught you Copernicus?'

'Magister Mästlin.'

'He should know better.'

'He refused to teach it in class because it went against Holy Writ but he showed me the ideas late one afternoon, when most students were outside catching the last of the sunshine. From the moment I heard the idea, I knew it was right. The Sun is too powerful to be anywhere but at the centre of creation.'

'It's unseasonably hot, don't you find?'

Kepler was finding it hard not to shiver in the draughty room.

Tycho downed another goblet and ran his hands over his face. 'I was sixteen – used to sleep with a cross-staff under my pillow so that I could observe when my tutor was asleep. He didn't approve of astronomy, thought it unfitting for a noble, but the thrill I felt when I pointed that simple instrument at the stars . . . I would spend half the night sliding the crossbeam to and fro along the staff, making it fit between pairs of stars and then reading the angles off the carved scale. One night I was taken by surprise at how close to each other Jupiter and Saturn were in the sky. When I checked Copernicus's book for his prediction of the conjunction, he was wrong. Ptolemy was more accurate – still wrong, but more accurate than Copernicus. It told me that neither was right, and I set out to correct things once and for all. Why is it so damned hot in here?'

He rose unsteadily to his feet and headed for the window. As he fought with the catch, a tiny breath escaped him. It was all the warning he gave before collapsing. As he fell, he knocked a tall candelabrum to the floor.

Kepler was at his side at once, rolling the great bulk over. There was a bloody bruise on Tycho's forehead where he had collided with the windowsill on the way down. His eyes remained open for a moment, locking with Kepler's, then flickered shut.

One by one, guests packed and departed, taking with them the lifeblood of Tycho's household. As the days went by, members of the family within reasonable travelling distance replaced them. Initially Jepp had taken to wailing outside the Master's bedchamber, but a well-placed boot from Tengnagel had stopped the dwarf from trying that again. With just the family and the assistants, and Jepp now keeping a mercifully low profile, the place began to feel ghostly.

Kepler worked on as best he could. Desperate for any news, he relived the scene over and over, wondering what more he could have done. That pitiful last look in Tycho's eyes haunted him, as did the dead weight of his Master's body when he had tried to move him and his strangled voice when he had cried for help.

Tengnagel had appeared almost at once, muscling Kepler aside with sword drawn and sending him crashing into the plaster wall. 'What have you done to my father-in-law?'

Kepler could not speak for fear, his eyes transfixed on the point of Tengnagel's blade. In his peripheral vision, he had been aware of Tycho's head lolling on the floor, the great man's legs twisted under that mighty body. He had to do something but what?

Thankfully Tycho's wife had arrived. Her scream drove away the murderous look in Tengnagel's eyes, and he had sheathed his sword and immediately attended to his father-in-law. Yet not even he could lift the unconscious form. He struggled for a moment, growing red as a beet, before grabbing the armpits and heaving. When Kepler made for the feet, Tengnagel had growled at him. Kepler's last view of Tycho had been of him being dragged away like a sack of firewood.

Occasionally snippets of information would come from upstairs; none were ever good. By all accounts, Tycho was in the grip of delirium.

'It's all he says, over and over in Latin: *ne frustra vixisse videar*,' came one report.

'Let me not seem to have lived in vain,' translated Kepler, a dark hole of foreboding opening within him.

One day, Tengnagel came to find him. 'He's asking for you.'

Kepler was fearful as he entered Tycho's private chamber. Before going in, he brushed the flecks from his jerkin and smoothed his hair.

The room was shuttered. Only the odd shaft of daylight found its way through a knothole or a split in the wood. Tycho's family were gathered around the bed. As Kepler's eyes adjusted to the dark, so he made out their haggard expressions and suspicious eyes. The room stank, and Elisabeth held a nosegay of dried lavender. Someone had placed a picture of the Lord on the mantelpiece at the foot of the bed. Tengnagel bent to his father-in-law's ear. 'It's Kepler.'

Tycho's eyes opened a crack, releasing beads of sticky moisture. The dying man beckoned Kepler to come closer.

'Tengnagel will take charge of my observatory equipment and observations. He is the only one of my family who has any idea what do with it. But you must promise me that you will complete my task, and publish the tables according to my system of the planets. Promise me.'

'Sir, I promise you that I will act only in the most noble of ways with your legacy. I will find only the most elegant solution.'

'Then you will follow my system.'

'I will follow the observations.'

Tycho's eyes filled with pleading but his mouth twitched. Kepler knew he had made him angry. Then the spasms stopped.

'Tyge?' His wife rose from her seat to lean over him. 'Tyge!'

Tengnagel pulled Kepler out of the way. 'Fetch the doctor.'

'It's too late for that,' said Tycho's widow.

Twelve imperial guards flanked Tycho's great coffin on its final journey to Prague's Old Town. Preceded by Tycho's coat of arms and his favourite horse, they led the long cortège of mourners across the city bridge, under the astronomical clock and across the market square to the Church of Our Lady Before Týn. The building's two gigantic towers reached up to Heaven.

Onlookers crowded the streets in silent tribute, brought out by the spectacle, even though many of them had never heard of the man.

Immediately behind the coffin was Tycho's weeping widow, supported by their eldest son. Then, of course, there was Tengnagel. Walking with his chin thrust upwards, he was glorying in the weight that was now upon his shoulders. At his side walked his wife, Elisabeth, quiet and composed.

Next came the nobles and gentry. Von Wackenfels was there, representing the Emperor. Kepler walked along behind them, in among the ranks of colleagues and collaborators, who included Jessenius. Bringing up the rear were the assistants.

The mourners packed the church shoulder-to-shoulder and listened to a torrent of unending praise for the astronomer. Everyone, it seemed, wanted to pay their respects. Jessenius spoke last, concluding the eulogies. He reminded the congregation that, while money and power die, art and science endure.

Afterwards at the Golden Griffin a feast of appropriately Tychonic proportions had been laid on. As the eating and drinking progressed, so the level of conversation rose, the odd laugh was heard, and soon it resembled any other party.

Kepler was seated at a table with Jessenius and von Wackenfels, content just to listen and observe.

'Thank heavens for the Utraquists and their liberal Catholicism,' von Wackenfels was saying, draining his goblet. 'Who here thought we'd say that?'

Jessenius nodded. 'It must have been difficult to decide how to send him off.'

'He hadn't been to church for twenty years. He wasn't a Catholic,' said von Wackenfels.

'Indeed not, especially as he spent some of his student days at Wittenberg.'

'He certainly couldn't be buried a Lutheran while being afforded the pomp that the Emperor demanded. And especially not with the ban coming.'

Kepler looked up. 'What ban?'

An embarrassed silence fell over the table.

'What ban?'

Von Wackenfels squirmed. He put his elbows on the table and spoke into his clasped hands. 'The Emperor is coming under increasing pressure from Rome. He must be seen to act. A decree is being drawn up to ban Lutheran practices.'

'Why did you not tell me before? I will be forced out.'

'Calm, stay your panic.' Von Wackenfels held up a palm. 'To be Lutheran is not outlawed, only the services.'

'And you of all people will be safe,' said Jessenius. 'It is widely known that you have your own issues with Lutheran doctrine: the Formula of Concord . . .'

'How do you know of that?'

'Father Grienberger, the Jesuit, has been visiting Prague. You know each other, I believe.'

'We have corresponded, yes. It was he who put me in contact with Hewart von Hohenburg.'

'He spoke very highly of you.'

But Kepler was no longer listening. All of a sudden there was a pain in his stomach. It was as though something in the meal had disagreed with him.

Back in his study in the little house on Karlova Street, Kepler found it hard to keep his thoughts on Mars. A fuzz of panicky questions constantly demanded his attention. Where could he escape to, if things became fraught here? Who needed a mathematician? Even if he could find another job, how could Barbara move in her condition? Maybe once the baby had been safely delivered he could think again.

The estuarine smell of boiled turtles coiled under the study door, threatening to make him retch. It reminded him of the hateful day he had screamed at Barbara.

They had still been in Graz. It was after a snowstorm, the very night that Heinrich, their first child together, had been born. The concern on the faces of the women in the delivery party should have told him something was wrong, but the baby's cries persuaded him that his worries were unfounded.

'A son!' he shouted as loudly as he could.

With the words ringing in his head, he rushed to his study to consult the planetary tables and chart his son's nativity. It was the second of February, 1598, and the constellations were promising something propitious indeed. Kepler noted with growing excitement that the moon was in quadrature with Saturn.

A noble disposition, a strong body, strong fingers and agile hands, with a capacity for mathematical and mechanical arts, he wrote down. As he further contemplated the celestial alignments, he could see the portent of a vivid imagination, compassion, piety, perhaps a hint of stinginess and mistrust – but with Europe in such turmoil they were probably good traits to have.

He bounded up the stairs waving his notebook in triumph.

Barbara was sitting up in bed, rocking the child. Tears had streaked her face. Little wonder, thought Kepler. At times the house had echoed with screams more akin to taking a life than bringing one into the world.

He was halfway through his proclamation when she said his name sharply, capturing his attention. She continued more quietly: 'Johannes, there is something you must see.'

With heart pounding, he watched her unwrap the baby. Lodged between the boy's legs was a peculiar carapace of skin, dark and craggy, where his testicles should have been. It looked like the gelatinous mass of a boiled turtle.

Kepler stared at the deformed genitals. He could think of only one thing: the liquor dripping from Barbara's smiling chin as she wolfed down another turtle. His eyes began to play tricks, doubling the image of his son and then quadrupling it. Dizziness and nausea assaulted him.

'Husband?'

He began talking – then shouting – and then screaming at her about how her gluttony had engineered this calamity. It all made perfect sense in his pain-drenched mind. The women began to wail, some running from the room with their arms thrown upwards. Barbara too began to weep, but silently.

The sight stopped Kepler mid sentence and, with a panic-stricken glance around him, he fled the room.

Later, his thoughts black with anguish, he returned to see Barbara. He sank to his knees and begged for her forgiveness at her bedside. 'I am worse than a mad dog, barking at those I love the most,' he pleaded.

She told him the matter was closed, but his shame would not budge so easily, especially when two months later Heinrich's final tragedy had overwhelmed them all. His tiny body had been consumed by an unquenchable fever that took him in a matter of hours. Worse, the same had happened a year and a half later to their next child, Susanna.

If the smell reminded Barbara of that awful confrontation then she gave no indication of it when Kepler joined her at the table. She occasionally looked up and smiled as her jaws mashed the softened cartilage.

Kepler could bring himself to eat no more than a few spoonfuls and excused himself. As he left the room he glimpsed Barbara reaching for his plate.

Later that evening Barbara fetched him from the study. Von Wackenfels had arrived and was pacing excitedly across the front room. 'I have the news you crave,' he said. 'His Majesty wishes to appoint you imperial mathematician.'

Kepler remained expressionless, not daring to hope. 'He has remembered that I am Lutheran?'

Von Wackenfels nodded emphatically. 'Yes, yes. You'll be safe under his protection. He wishes to meet you tomorrow. Be at the Palace at nine.'

The front door had barely closed when Barbara flung her arms around Kepler, her weight throwing him against the oak. 'Husband, you are my prince.'

Kepler followed von Wackenfels through the grand hall milling with courtiers, and then into an anteroom decorated with calendrical murals. Above the harvests and frosts, storms and flowers, the depictions turned to the base elements of nature: a green field and mountains to represent earth; a ship on the stormy ocean for water; a twisting pillar of flame for fire; and an extra-vagant cloudscape with eagles for air. The god Jupiter represented the celestial realm. His conquering eyes regarded Kepler from above the inner doorway as if determining whether to let him pass.

Von Wackenfels drew open the far door and Kepler walked into a large chamber. He could scarcely believe such grandeur existed. Pictures were on the floor, resting against the walls, or up on benches in an approximation of where they might hang.

Most featured elongated characters in stylised postures. Many of the paintings seemed to be at least partly mythological in subject matter. Kepler recognised Urania, the muse of astronomy. The cherub was holding a sextant and dancing.

There were a number of bare-breasted women suckling babies. The mothers all possessed exaggerated necks so as to look down on their infants.

'This gallery will be one of the wonders of the world when it is finished,' said Kepler, wishing he could linger.

Von Wackenfels looked puzzled.

'The paintings, when they are hung,' said Kepler.

'But His Majesty prefers them this way.'

Kepler decided to keep his thoughts to himself from then on. He could ill afford a blunder like that in front of Rudolph.

From the gallery they passed into another similarly sized hall. This one housed exquisite mechanisms. Some of them Kepler recognised as astronomical instruments; others he could not even guess at. Suits of armour, some dented or split from battle, shared the floor space with display cases crammed full of glittering cups and amulets. From the ceilings the skeletons of animals hung down like macabre puppets. Some of them were so outlandish that Kepler could not even begin to imagine what they had looked like in life.

'We call this the Kunstkammer, His Majesty's Chamber of the Arts.'

'I have never seen such a place.'

'Nothing like it has ever existed before. It is the finest collection of human knowledge and artefacts ever amassed. Beyond are cupboards of manuscripts and other relics. His Majesty has everything in here from the latest philosophical books to unicorn horns.'

'I can see why he is loath to leave these rooms. Sorting through all this could keep a man fascinated for a lifetime.'

Von Wackenfels led Kepler up a steep staircase and out into the daylight. It was a glorious day, the sky powder blue; the kind of day that suggested the trees had been tricked into shedding their leaves too early.

They were on the top of a tower. The Emperor stood by the ramparts, looking out over Prague. From this distance, the city's buildings looked like toys, and Rudolph as though he could reach out and rearrange them. Above the tiny red roofs, starlings were massing in preparation for their annual migration south.

'Johannes Kepler, Your Majesty,' announced von Wackenfels.

The darkly clad figure did not turn round. Von Wackenfels jerked his head, indicating that Kepler should approach.

'Your Majesty,' Kepler said with a bow, 'I am here as your humble servant.'

Rudolph muttered something that was lost to the wind. Kepler edged closer.

'See the bridge?'

From this distance, its exquisite detail was lost. It was nothing but a strip of road that crossed numerous stone pontoons. Underneath slid the water of the Vltava, snaking crests of white foam marking its progress.

'Yes, Your Majesty.'

'My ancestor Charles IV founded that bridge. He did it at the very moment when the Sun eclipsed Saturn to shield it from the planet's evil influence.'

'His Majesty's forefathers were wise indeed.'

'Astrology is of the highest importance here. We rely on it to guide us. Tycho understood that.'

'And I will endeavour to succeed him in every way that I can, Your Majesty.'

Rudolph made a small sound. 'There is someone I want you to meet.'

Kepler followed in silence back inside and downstairs, feeling increasingly uncomfortable. At one point, he snatched a glance at von Wackenfels, a few paces behind. The councillor smiled reassuringly, and Kepler relaxed.

When Rudolph did speak, Kepler strained to hear above their footsteps.

'We will buy Tycho's equipment and observations from his heirs. We've made them an offer of twenty thousand thaler. You will do everything you must to publish *The Rudolphine Tables*, as Tycho promised.'

'I will, Your Majesty.'

'You will be paid for this work by the Palace.'

'Most generous, Your Majesty. May I respectfully discuss the salary? My wife is expecting a ch—'

Rudolph stopped him with a squeak. 'The Privy Councillor will deal with that.'

Instead of heading into the Kunstkammer, they passed through a doorway and out into a sloping garden. Full sunshine greeted them, and Kepler had to raise a hand to shield his eyes.

'Welcome to the Garden of Paradise,' mumbled Rudolph.

It was a large triangular space lain mostly to lawn, interrupted by the occasional pruned shrub. Exotic squawks such as Kepler had never heard before came from a stone building with a chimney that twisted ribbons of smoke into the air.

'My aviary,' said Rudolph. 'We have to heat it, otherwise they die. Such a waste after all the trouble it takes to bring them here.'

A curious creature, resembling a black duck in shape, but around four times the size, waddled up to Kepler. Its large beak clicked open and closed, and its tail feathers curled like the plumage on some of the hats Kepler had seen at court that morning. He tried to sidestep the bird. It flapped its vestigial wings as if trying to work out what they were for. Then it turned to intercept him again.

'There is nothing to fear. It is quite harmless,' said Rudolph. 'They call it a dodo because it is utterly stupid.'

Von Wackenfels shooed it away when Rudolph was not looking. Kepler mouthed his thanks and hurried to catch up.

The Emperor led them to another stone outhouse, much larger than the first. At the doorway, Kepler heard unmistakable growls from within and hesitated. Rudolph giggled and walked on. As Kepler followed, the tang of animals caught in his nostrils.

The floor was covered in straw, and animals were chained along the walls. A prostrate wolf pricked its ears and looked up at the new arrivals. In the next stall, a black bear rocked its head to and fro, paying them no heed. Kepler stared in amazement at the next animal. It was a tiger. The vividness of its coat, the burnt orange and black stripes, entranced Kepler as did the lazy way it lifted its head and flicked its tail.

Arriving at the last stall, noticeably finer than all the others, Rudolph stopped. 'And this is my lion.'

The beast was lapping at the eviscerated ribcage of a deer. The stench of blood hung in the air. The lion was so caught up in its meal that it completely ignored the humans. Its fur was the colour of expensive honey but its muzzle was stained red. It shook its head, rippling its luxuriant mane, and then returned to its meal.

'Tycho told me we shared similar horoscopes.'

It took Kepler a moment to convince himself that he had heard correctly. He was about to object to the idea of an animal being in any way similar to a human.

'I believe it to be true. We are brothers.' Rudolph gazed lovingly at the animal.

Kepler looked away lest his face betray shock. *Animals have no souls; how can the Emperor be so naive?* As Kepler's eyes sought something else to rest on, he could not help but notice there was enough meat left on the deer's carcass to feed his own family for a week.

14

The next summer, as God walked among his people in Prague, so the Devil was but a half step behind. The burning wind withered crops and parched the city of moisture, clotting the air into a breathless mass that clung to the streets as molasses to a barrel. The sun glared, cracking the stucco façades of the buildings and blistering the faces of the inhabitants.

As tempers frayed, so rumours of plague in Hungary caught hold. Each stricken body was reputed to display bloody marks corresponding to where Christ had been nailed to the cross. This was taken as proof of the divine retribution being meted out. Undertakers in Prague searched the recently deceased for such marks; officials waited anxiously for news of the plague reaching the city.

In this cauldron, the Keplers prepared for their third baby. The shutters at Barbara's window prevented light from entering and heat from leaving. Regina spent much of her time in the simmering bedroom, dousing her mother with a wet cloth.

During Barbara's confinement, Kepler slept alone. Often he would rise in the night to scribble down some thought or to crank through some celestial calculation that had seemed intractable the day before when the closeness of the air had transformed his study into a furnace.

Today, however, he had entirely new things to worry about.

In the far corner of the coaching inn's courtyard, a black coach was being unloaded. He made his way through the crowd to where a young man was dropping the travelling cases down to a colleague.

That's when he saw her, smaller than he remembered and her shoulders now rounded, but unmistakably Katharina: she was chiding a boy whose only crime was to help with her luggage.

'Careful with that,' she said.

The boy swung the bag up on his shoulder. 'Just tell me where you want to go.'

Kepler rushed over. 'I'll take it from here.' He handed the boy a coin.

'You're welcome to her.' The boy dumped the bag on the ground, gave her a sour look and sauntered off.

'You're not getting a tip from me,' she called to his receding back.

The boy raised a hand without looking round and moved his fingers and thumb to imitate a flapping duck bill.

Kepler turned to the woman. 'Hello, mother.'

'You let him talk to me like that.'

'Mother, it's of no concern. Here, let me take your bag and escort you home.'

Her bad mood did not last long. It seldom ever did. The people and their clothes soon enthralled her; the neat cut of the summer jackets and the colour of the dresses. They pushed through the narrow passageways, dwarfed by huge churches and municipal buildings, occasionally coming to a crossroads or a square where they could glimpse the sprawl of the city around them.

'My heavens, it's a wonder,' she said, bending backwards to look from the top of one building to another and another. 'I never dreamed it could be so big.'

'From the Palace up on the hill, the city is a beautiful sight. You look down on the rooftops.'

She took his arm. 'I still think of you when you were five, running behind the bar, totting up the rounds in your head. People used to come to the inn just to watch you.'

Kepler felt himself blush though he could not say why. Those had been happy years at his grandparents' tavern, despite his father's unpredictable comings and goings.

'Now look at you, Mathematician to the most powerful man in Europe. I can scarcely believe it.'

'Nor I, Mother, nor I.'

Kepler carried the case up the stairs to the attic and showed his mother to her room. The floorboards had not been polished up here; a small dormer window allowed in some light.

'Opposite the servant's room . . .' she said.

Kepler took her back down the stairs and into his temporary bedroom.

'I'll have Frau Bezold change the sheets.'

'You don't need to do that; we're family.' Katharina was suddenly magnanimous again.

'Nevertheless, you're my guest, Mother. I'd like you to be comfortable.'

'Well, I'll do it then, no need to make a fuss.'

That evening, Barbara started sobbing. Her muffled sounds carried through the house to where Kepler and his mother were sitting. The heat was insufferable, and his shirt was soaked at the armpits. 'She grows more fearful with each passing day. Nothing I say comforts her, but she will quieten near morning when Venus rises.'

'What's she scared of?'

'That we'll lose another child. Barbara is a pious woman. She is a beautiful woman, but she is no longer a strong woman. She did not deserve to lose two babies. I fear a third bereavement might unbalance her completely.'

'You have to put your trust in God.'

Kepler nodded. 'Why must life be so harsh?'

'We all waver in our belief at times, even me.'

'They were so young; neither of them made it past their second month. I see them still in my mind. I notice children in the street that are of the age they would be now. What can God want with children that small?'

'What if He saw something terrible in their future and decided to spare them?'

Kepler opened his mouth to speak, thought better of it, then plunged on anyway. 'Then why didn't he weed out my father? His fighting did none of us any good. Especially not when he turned his fists on you.'

Her eyes narrowed into beads. 'Because then you would not have been born.'

From upstairs, Barbara's sobs reached a crescendo. Kepler tried to shut his ears to them.

'She cannot carry on like this,' said Katharina.

'The doctor says there's nothing he can do,' Kepler snapped. 'We must accept it.'

'You might,' she said, her face thoughtful.

Next morning, Katharina was nowhere to be found. Kepler checked her room, releasing a pent-up breath when he saw that her luggage was still there. He bounded downstairs to the scullery.

'Do you know where my mother went?' he asked Frau Bezold.

'Out, that's all I know.'

Kepler was still pacing when the latch went on the front door. Katharina walked in carrying bunches of wild flowers and leaves. 'Found some,' she said, pointing to a number of stems, each a foot or more long, with pale green leaves and clusters of tiny pink flowers close to the stem. 'They're not too strong but better than nothing.'

'Mother, this is not Leonberg, you must be careful in the streets.'

'I've been out to the woods.'

'There are thieves everywhere.'

'Look at me, Son. Do I look as though I have anything worth stealing?' Her tiny frame was covered in a grey country dress, and she wore clogs on her feet. 'Most people look at me as if I've wandered into the city by mistake.'

'Then think about how it looks to people who don't know you: an old woman collecting herbs for potions. You would do well to remember what happened to your aunt . . .'

Katharina silenced him with a look.

'Sorry, Mother. It's just . . .'

'I know. Don't be afraid. That was a long time ago.' She reached up to his shoulder. 'I need your kitchen.'

She did not wait for his answer before heading in its direction. Kepler trailed after her, to smooth things over with Frau Bezold.

Katharina tied her stems into bunches and hung them in front of the window, where they could catch the strongest sunlight. 'I'll be back later.'

Kepler and Bezold exchanged shrugs.

In the evening, Katharina returned to the kitchen to complete her preparation. Before long she was carrying a bowl of liquid upstairs to Barbara. Its bitter steam reached Kepler's nose.

'Are you sure this is wise, Mother?'

'Trust me. It's only motherwort and a few other things.'

Barbara was sitting on the edge of the bed. She was cupping her heavy abdomen, gently stroking her thumbs up and down. Regina stood next to her, massaging between her mother's shoulderblades.

'Here you go, drink this.' Katharina held the bowl so Barbara could see it. The pregnant woman's eyes flitted to her husband.

He nodded.

When the bowl was inches from her mouth, she pulled a face. 'It smells bitter.'

'It'll do you good,' urged Katharina.

Pinching her nose, Barbara sipped at the potion, taking bigger and bigger mouthfuls until it was all gone.

'Now then, let's get you back into bed,' said Katharina, taking the empty bowl. 'Help me, Regina.'

Together they swung Barbara's legs back onto the mattress, as she heaved herself up on her arms.

That night, there were no sobs. The night after, Kepler heard Barbara laughing again.

On the night of the birth, Kepler was buffeted by the agonising screams coming from upstairs and the constant bustle of women fetching cloths and water. Adrift, he wandered from room to room but nowhere could he find any peace of mind, not even in his study amid the piles of Tycho's ledgers, which he had appropriated after his appointment by the Emperor.

His suffering was finally broken in the early hours when the newborn's cry echoed through the house. As he climbed the stairs, he became suddenly fearful of what might await him, and slowed his step.

When he reached the bedroom, he tiptoed inside. The baby had quietened and was slumbering in Barbara's fleshy arms. His wife's round face was haggard, the first time Kepler could remember thinking of her in that way, and at the sight of him she started to cry.

Regina delivered the news. 'I have a sister,' she told her stepfather with a delighted hug, her head resting just above his stomach.

'Is she . . .?' Kepler's voice caught.

Barbara held the baby for his inspection. 'Husband, she is perfect.'

He took the miniature form into his hands and stared into her face. Her eyes were tightly shut and her lips were pressed together. Her tiny hands bobbed. It was then that he understood that his wife's tears were those of joy.

'Let us call her Susanna,' he said.

'Dare we use that name again?'

'She will be our constant reminder of God's divine wisdom.'

A few days later, a sharp rap on the front door sprang Kepler from an all too brief sleep. He felt as if the strength had been baked from him during the night. Susanna's cries for food and

Barbara's complaints over the pain of suckling had not helped either.

More ladies to welcome Susanna to the world.

He straightened his jacket in preparation for playing host but upon entering the front room, he saw that it was Jessenius. Silhouetted at the window was a towering figure. The mountain of black fabric turned slowly.

'I believe you two have corresponded with each other,' said Jessenius. 'This is Father Grienberger.'

'Father Grienberger! My apologies, no one has bade you sit down. Please excuse my housemaid, she's old and weak of mind. We must . . .'

'It is of no concern. We are here to speak with you as friends, not to judge you on etiquette.'

'My wife gave birth some days ago. We are still getting used to this joyous adjustment in our lives.'

Grienberger smiled, an expression that did not sit comfortably beneath his enigmatic blue eyes. 'We share your good news.'

'It is why we're here,' Jessenius added, earning him a warning frown from Grienberger.

Alerted, Kepler stared at the Jesuit. The moment threatened to become awkward but, thankfully, it was broken by Frau Bezold's entrance. She set down a tray laden with goblets and a pitcher of wine on the table, and the men gathered round.

Kepler served his guests with trembling hands and after some small talk about the heat, he asked, 'How may I be of service to you today, gentlemen?'

Grienberger inclined his head, not quite looking at Kepler. 'It is a delicate matter.'

'Are we not friends?' asked Kepler, watching Jessenius closely.

'We are concerned for your daughter,' said Grienberger. 'You are not known for fathering strong children and you know full well the peril to your daughter's soul, should she pass away without being baptised.'

Vertigo wheeled inside Kepler; he looked from Grienberger's averted gaze to Jessenius. He too was studying the walls.

'You are suggesting I baptise my daughter a Catholic.'

'Lutheran services are outlawed. It is the only way.'

'She is an innocent; her soul has nothing to fear.'

'You forget; we are all born with Adam's guilt on our shoulder.'

'You need only look at her to know that she is pure.'

'Even Luther believes in original sin.'

'Then Luther is wrong.'

The Jesuit cocked his head, drawing Kepler up sharply. A profound sorrow grew from nowhere to fill the astronomer completely. 'I cannot baptise her a Roman Catholic.'

'We would organise something quiet for you. No one need know.'

'I cannot. I have taken the decision. Since my daughter cannot be Lutheran, she will be baptised Utraquist.' The decision had cost him nights of sleep. He had contemplated smuggling the family out to one of the surrounding castles where it was said that Lutheran ministers were hiding. However, if he were to be caught he would certainly be replaced as imperial mathematician, and he could not risk another expulsion like Graz, not with a newborn. A Catholic baptism was out of the question; there was no way he could yield to the Pontiff; it would be an unthinkable betrayal of his upbringing and education, not to mention his personal conviction. God had given men minds to use, not to surrender them to blind obedience. To do so would be the equivalent of a farmhand refusing to lift a scythe. The only alternative was the Utraquists. At least by maintaining their own identity they had demonstrated some ability to stand up to Roman pressure.

Kepler rose from his seat. 'I thank you for your concern and I bid you both a good day.'

Their exit was conducted in silence. As Kepler returned to the foot of the stairs, wondering what to tell Barbara about the encounter, he heard his mother's indignant voice coming from the kitchen. From the sound of it, Frau Bezold was rising to the challenge.

The two women were engaged in a tug of war over the remains of a shoulder of mutton. Some remnants of meat were all that hung from the bone.

'It's spoiled in the heat,' Frau Bezold was saying.

'You can cut that off.'

Kepler raised his voice. 'Mother, what are you doing?'

She did not let go of the bone. 'She's going to waste this.'

'It's gone off,' said Bezold.

'Enough, the pair of you! Give me the bone.'

Kepler held out his hand. After a moment, Bezold relinquished her grip and Katharina passed it to her son. Sure enough there was a green tinge to the scraps that remained.

'This joint is spoiled. We cannot eat it.'

'You were not brought up to live a wasteful life,' accused Katharina.

'Oh, mother, if only you knew how tightly our belts are tied. It is an effort just to get the salary that I was promised from the Palace. I spend most of my time chasing from one office to another, trying to find someone who won't fob me off saying it isn't their concern.'

'But the Emperor said he would pay you.'

'Yes, but the actual business of handing over the money is done by officials who are overwhelmed by the demands of their master.'

'Then I will go up there tomorrow. I'll make them pay up what they owe you.'

'No, mother, you won't. I can fight my own battles.'

Crossing the bridge one morning, Kepler stopped to mop his brow and noticed six or seven feet of cracked mud beneath. The Vltava usually sloshed all the way up to the stone embankment. He had to stop several more times on his walk up Hradčany

Hill. His best doublet and jerkin were too heavy for this weather, but what else could he wear to the Palace?

A red-faced von Wackenfels was waiting for him, his blond hair newly cropped.

'You've cut your hair.'

'Vanity gave way to practicality.' He ushered Kepler into the Palace and guided him to the grand hall, which was full of gentlemen fanning themselves with their hats.

Kepler approached the vast windows in the great hall, thrown open to allow in what currently passed for fresh air, but the heat from the stone columns drove him back.

'You look tired, my friend,' said von Wackenfels. 'Are the demands of fatherhood testing your resolve?'

'It's this constant business of visits. I had no idea my wife was so widely known in Prague. She has amassed a veritable army of friends. Each one must be treated with such cordiality on arrival and exit that it drains my time.'

'Nevertheless, you have completed the prognostications?' Von Wackenfels sounded anxious.

'Indeed, I have.' Kepler patted the documents, tied into a sheaf. 'It is *The Rudolphine Tables* that are falling behind. But I'll catch up.' He passed the documents to von Wackenfels.

'I have another issue to discuss,' said Kepler. 'My salary is now three months in arrears.'

Von Wackenfels grimaced. He leaned in and lowered his voice. 'I am sorry, my friend. These are difficult times for the Emperor's finances. We are pressed from all sides. The presentation of this should help though.' He waved the horoscope. 'Nothing loosens the imperial purse more than a gift.'

'Is there nothing that can be done today?'

Von Wackenfels considered the problem for a moment. 'Listen, I have a surfeit of wine. I'll have some brought round to you.'

'I don't want charity.'

'It isn't charity. It's a gift while I see what I can do on your behalf. You must drink in this heat; my wine will help.'

Kepler nodded his thanks.

'You're not the only one who hasn't been paid.' Von Wackenfels tipped his eyes across the room.

Striding towards them from the inner chamber was Tengnagel and a number of hangers-on.

'What's he doing here?' whispered Kepler.

'He is an Appellate Councillor now.'

Tengnagel's heels clicked on the flagstones. As he drew close, Kepler noticed that a crucifix was bound around his belt. Flanked by his cronies, the Junker stopped, raised his chin and looked down his long nose.

'Herr Kepler.' He sounded cross.

'Junker Tengnagel. You have exchanged astronomy for politics.'

'Where are the great and illustrious observations of my father-in-law?'

'In my study, at home, where I can work on them.'

'And who gave you permission to remove them from the Golden Griffin?'

'The Emperor purchased them from you and commanded me to produce the tables.'

'The Emperor has not paid me so they are still mine. That means you stole them from me.'

Von Wackenfels stepped in. 'You have been paid some money.'

'A thousand thaler, which is no more than the interest on the debt. I want my father-in-law's observations back.'

'I cannot go against an imperial command. I will not hand them over.'

Tengnagel smirked. 'Then I propose a deal, since you are using my observations. Everything you do with them must be published under our joint names.'

Kepler stilled his first instinct, thought about it, and then spoke: 'I agree, on one condition: that you pay me a quarter of all interest monies you receive for the observations. Two hundred and fifty thaler per annum – a small price to pay for immortality, wouldn't you say?'

Tengnagel's brow knotted. He pointed at Kepler. 'Your impudent greed knows no bounds. You'll get nothing from me. Prepare the ledgers. Someone will be round to collect them this afternoon.'

Kepler was left standing open-mouthed, unsure how this disaster could so suddenly have befallen him.

'I'm sure we can do something,' said von Wackenfels.

Kepler was not listening. 'I have to go.'

'Johannes, wait!'

'I have no time,' he shouted over his shoulder, drawing stares from the others as he sped out of the hall.

Von Wackenfels caught up. 'At least let me pay for a carriage for you.'

When Kepler burst through the front door, Barbara was nursing Susanna in the coolness of the shadows. She jumped from her seat. 'Husband! You gave me a start.'

'Take Susanna upstairs. I have no time to lose.'

As she scurried away, he separated the Mars ledgers from the others and laid them in a stack in strict chronological order. They comprised a tiny fraction of the whole collection. If he were lucky, Tengnagel's men wouldn't notice they were missing. He checked to make sure he had them all and then carried them five at a time to the attic, setting them on the floor of his temporary bedroom.

On one trip Katharina appeared. 'What's all this running up and down?'

'I have no time to explain, mother. Please, stay out of sight.'

She shrank back into her bedroom.

When the last trip was completed, he heaved the mattress off the bed and laid the books in a single layer across the wooden frame. Eyes stinging with sweat, he pulled the mattress back over them and rearranged the covers.

He finished just as the bailiffs arrived, instantly recognisable by their rhythmic thumping on the front door.

15

1604

The carriage rumbled through the night, bumping across the Stone Bridge. Inside, Kepler tried to rub some life into his eyes, which not ten minutes ago had been shut tightly in sleep. As well as playing havoc with his rest, these late night summonses irked Barbara, and with good reason, thought Kepler. They woke Susanna who then toddled around the house, behaving as if it were morning.

The Emperor demanded more and more attention these days, commanding reports or horoscopes whenever a whim entered his head. Lately it had been the nativity chart of Caesar Augustus; then one for Mohammed; followed by the fate – according to the stars – of the Turkish kingdom; and a prophesy on the outcome of a fight between the Republic of Venice and the Pope. The list was never-ending.

Kepler sniffed loudly, trying to clear his nose. He swallowed down the foul-tasting stuff and yawned as the carriage started up the hill to the Palace. When they reached the courtyard, a bleary-eyed footman opened the door.

Von Wackenfels was waiting, shifting from one foot to the other.

'What *is* the urgency?' asked Kepler, dismounting.

'You don't know? There's a new star in the sky.'

Kepler looked up. It was thick with cloud.

'The Imperial Meteorologist reported it earlier, before it became overcast.'

'A meteorologist?' Kepler struggled to keep the sarcasm out of his voice. He knew he was being uncharitable but, at this time of night, he found being grumpy helped, especially if he thought he had been woken for nothing.

Von Wackenfels frowned at him. 'He's no fool. He and the Emperor are up there now, waiting for it to clear.'

As they hurried along one black corridor after another, Kepler finally recognised the mural of the god Jupiter above the entrance to Rudolph's Chamber of Art. They hastened inside.

'Johannes, you should know that the Emperor is not himself these days. He has a lot on his mind. Having not married, he leaves no heir.'

'He's only fifty-two.'

'Nevertheless, he feels time is running short.'

'Then he will be succeeded by his brother.'

'Yes, Matthias is the obvious choice. He is well regarded by the rest of the family and has a fine military mind, but he has been so antagonistic towards His Majesty that the Emperor would prefer to be succeeded by his cousin.' Von Wackenfels hesitated before saying the name. 'Archduke Ferdinand of Styria.'

Kepler's insides churned. 'The man who exiled me from Graz.'

Von Wackenfels nodded.

They climbed the tower and returned to the night air.

Rudolph was dressed in little more than his nightshirt and a gown. He looked thin and was tottering on tiptoes, moving from one parapet to another, anxiously scanning the sky. His hair was unruly. He mumbled something as von Wackenfels announced Kepler's arrival.

The Meteorologist was a young man with an earnest face. He wore a long cloak and stayed fixed in his position, watching the voluminous clouds. 'It was in this part of the sky, brilliant bright and sparkling with colour,' he said to Kepler.

'Are you sure it wasn't Jupiter? It's bright, though it doesn't twinkle.'

'It was close to Jupiter, sir, but it was not that planet.' He radiated sincerity.

They waited together as Rudolph flitted around behind them, saying something in Latin that Kepler could not catch. The rest of them passed few words. After an hour, an unmistakable dampness rose in the air.

'If it turns misty, we are finished tonight,' said Kepler.

'There!' The meteorologist was leaning over the edge, pointing to the sky. 'It's clearing.'

The clouds were shrinking from the sky so quickly that it could have been the hand of God brushing them aside. More and more of the sky opened up. No one spoke; they just stared and waited. Even Rudolph paused in his strange dance.

'Oh my . . .' said Kepler as a quartet of jewels was revealed. He identified three of them immediately: red Mars, ochre Saturn and white Jupiter, gathering for their great conjunction. The fourth took him completely by surprise. Unlike anything he had seen before, it was certainly not a comet because it had no tail. It rivalled Jupiter in brightness, but in contrast to that steady beacon the new star glittered, displaying all the colours of the rainbow.

Rudolph sank to his knees to receive the light of the new star, arms open and palms upright. His voice rose in volume and clarity. 'Of all the times for this new star to appear, it chose this special moment. What are we to make of it, Mr Stargazer?'

'It is propitious indeed, Your Majesty, though I cannot give you a snap assessment.'

'Forgive me, what special moment, Your Majesty?' asked the Meteorologist.

'We are entering a new astrological era, one that stands for righteousness,' explained Kepler. 'It happens only once every eight hundred years and is heralded by the grand conjunction of Jupiter and Saturn as we enter the fire trigon, the beginning of the cycle . . .'

Rudolph interrupted. 'In the past, the entry into the fire trigon has heralded the greatest turning points in history and the ascendancy of a great man to guide the way. Eight hundred years ago, Charlemagne founded Europe; eight hundred years before him Christ was born. The question is, what now? What great age are we about to enter and who is to bring about this new dominion of thought and leadership?' He paced the rooftop, scratching at his forehead. 'The conjunction of the planets must have ignited this new world, brought it into being. Imagine, we have seen the birth of a planet. What influence will it cast on us?'

'Your Majesty, if it is anything like the new star that Tycho Brahe observed in 1572, it will be located far beyond the planets, in the starry sphere of Heaven. I believe this is a new star rather than a new planet. I will arrange for observations to be taken each night. If it's a star, it will remain fixed in position with the other stars. If it's a planet, it will follow its own path, and we will see this motion in a matter of days. I confidently expect to find that it is a star and that it will fade, as the previous one did, over the course of a few years.'

'I was crowned in the same year as the previous new star. Do you suppose that dooms me to fade away too?'

'Astrological symbols are seldom explicit in their meaning, Your Majesty.'

Rudolph was not listening. 'The new age is coming, what shall I do with it? Prepare me a horoscope. I need to know what this all means. I want to use this moment to crush my enemies.'

'Enemies?' Kepler spoke carefully. 'Religious enemies?'

'No, my own brother, Matthias. I will put a stop to his schemes once and for all.'

PART II
Culmination

16

Prague, Bohemia

1610

Karlova Street was one of the busiest thoroughfares in Prague. It was the last place in the Old Town that people passed through before crossing the Stone Bridge into the New Town, where lay the hope of imperial favour. And it was the first place to drown your sorrows on the trudge back home if things did not work out.

There were a number of inns squeezed into what used to be homes but had been hastily converted when the bridge opened. They attracted a steady stream of customers for as long as they were open. For those on tighter schedules or budgets, a pair of women in heavy shawls sold mulled wine from an urn over a fire in the street. For a few coppers they would ladle the aromatic drink into dented mugs. Just down from them, in his usual spot, a bearded knife-sharpener pedalled his whetstone.

All day, people trooped to and fro. Only in the evenings did the hubbub show any sign of abating, replaced by the occasional coach and the passage of evening revellers. Kepler hardly noticed these comings and goings any more; after a decade in the city his brain had long since learned to filter out the sounds.

Barbara sat facing the empty fireplace. She was wearing an old green dress that had once been saved for best. Now, spoiled by a few pulled threads, it served as everyday wear. Nothing had replaced it for best. Her arms were drawn tight, her face impassive. Most ominously of all, her mouth was drawn into a tight line.

Kepler had seen this mood brewing for days and had avoided it thus far, taking refuge in his study. Yet, the frustration of the silences at mealtimes and her turned back in bed at night were growing impossible to ignore.

'You've not been yourself lately,' he stated matter-of-factly.

She ignored his comment.

'Is it the children?'

In addition to Susanna, who was now eight, six years ago Barbara had given him Friedrich, and four years later Ludwig. The household rang with their chatter and Barbara was constantly busy.

'Do they tire you out?' he asked. That at least brought a look of disdain. 'Then what is it?'

'Nothing.'

He edged into her line of sight. 'Perhaps we should move. There are houses available in the New Town, close to the Palace. There's a better class of people over there.'

Barbara's head jerked around. 'You don't understand, do you? We are objects of ridicule to the better class. "Mr and Mrs Stargazer", that's all we are to them – ridiculous.'

'It's not ridiculous to be known for what you do.'

'We don't have the money to move house. What good is being Imperial Mathematician if you're not paid?'

The comment might just as well have been a knife between his ribs. He certainly felt the same physical pain. If only she would give him credit for the hours he spent standing in queues, shuttling from one office to another, begging for his ever-accumulating wages.

'I sent a copy of the *Astronomia Nova* to England. I have no doubt that when the King reads it, he will look upon us with favour. We must just be patient, better times are ahead . . .'

'James of England now! When has a single one of your schemes to win favour ever worked for us? Not so long back you convinced me we were moving to Tübingen because of this great new book of yours. Look what happened there.'

Kepler winced at the memory.

He had arrived unannounced at the university, having taken a detour on his way back from the Frankfurt Book Fair. The building was exactly as he remembered, all draughty halls and high windows. Waiting in the ornate entrance hall, he was unable to shake an unsettling feeling of coming home.

A group of young men in short black gowns passed by, displaying that universal student trait of one moment practising airs and graces, the next being consumed by horseplay. As Kepler watched, they jostled each other while approaching an archway, engineering it so one of their number would end up colliding with the stone pillar.

Kepler arranged his collar and straightened his jerkin, recalling the time he had sewn up the arms in Zimmerman's gown, forcing the wastrel to arrive at lectures apparently armless. It may not have been the most original prank in the world but it had been amusing at the time. Even when Zimmerman correctly guessed the culprit and Kepler was marched to the Chancellor's office, where he confessed and took his punishment, it had still been worth it.

His memories transformed into the sight of Mästlin rushing down the corridor, long gown billowing. It brought a smile to Kepler's face to see him again. Time, it seemed, had stood still for the Magister, and, for a moment, Kepler could believe that he was a student again, waiting for a tutorial.

'Why are you here?' hissed the Magister, shattering the daydream.

'Am I not welcome at my old university?'

Mästlin avoided eye contact. 'Of course you are.'

'I wanted to see you, Magister, talk to you. I miss our discussions.'

'I'm a busy man, Johannes. You should have written ahead.'

'Magister, you haven't replied to my letters for more than five years.'

The side of Mästlin's mouth twitched. 'Johannes, your questions are beyond me. You have taken your mathematics to such a level that I cannot follow you any more.'

'But you have read my letters?'

'Yes, of course.'

'And did you read my book on the new star? I sent you a copy.'

Mästlin glanced around the hallway. 'You'd better come to my office.'

As Kepler followed, he noticed the scrutiny he was receiving from the fellow academics. He acknowledged one whom he recognised, but the old professor looked away.

'Is everything alright here?' he asked Mästlin.

'Yes, fine. Here we are.' Mästlin swept into an office, closed the door behind them and then looked perplexed about what to do next. He settled into his own chair, elbows resting on the desk.

'The new star stayed firm in its position until it faded away. It was clearly in the realm of the fixed stars, and so clearly at odds with Aristotle's claim that the cosmos is unchanging. The old cosmology is wrong, possibly to its very core.'

'Johannes, one small light in the sky cannot be used . . .'

'Not *one*, it is clearly similar to the new star Tycho saw decades before. And there is more, Magister, much more. I have simplified Copernicus's concept of the Sun-centred universe to make it so obvious, it has to be true.'

'Simplified it in what way?'

'I have fought a war with Mars these past years, and now I have him bound in the chains of computation.'

'Oh, do speak plainly. I remember this habit from your student days. It was annoying then.'

'Very well,' Kepler said stiffly. 'I have found the true shape of Mars's orbit. It is not a circle but an ellipse.' He might as well have said that God did not exist.

Mästlin looked at him with frank disbelief in his eyes.

'I can thread every single one of Tycho's Mars observations. I have removed the need for every epicycle, every equant, every deferent. No more whirling cogs to set the mind reeling. All you need is the single ellipse for each planet. Simple.'

'Where's the Sun in all this, in the middle of the ellipse?'

'No, at one of the focal points. Let me demonstrate. Do you have a length of string?'

Mästlin performed a long-winded search in the drawers of his desk until he produced a length of rough twine. Kepler knotted it into a loop. 'Now paper, pen and ink.'

Mästlin pushed a quill resting in an inkwell over to him, then retrieved a sheet of blank paper from a drawer.

'And I need your fingers.'

Mästlin looked puzzled.

Kepler adjusted the loop of string and placed it on the paper. 'In here, please.' He tapped his forefingers inside the loop, about three inches apart. Mästlin complied.

'Now, don't move them.' Kepler inked the nib, pulled the string into a triangle with Mästlin's fingers as the base, and hooked the quill into the apex. He then drew a complete loop on the paper, allowing the shifting tension on the string to guide his pen around and back to where it started.

'There,' he said.

Mästlin removed his fingers, and Kepler cleared away the string. On the paper was an ellipse. Kepler drew a dot where one of Mästlin's fingers had been and labelled it 'sol'.

'The distance between your fingers determines how extreme the elliptical shape becomes,' said Kepler. 'Closer together makes the ellipse more circular, further apart makes it more elongated.'

'If the Sun sits at one focus, what is at the other?'

'Nothing.'

'What guides the planets?'

'A force emanating from the Sun that becomes weaker with distance. I know that because I have found another law of planetary motion.' Kepler picked up the pen again and drew two diverging lines from the Sun to the furthest end of the ellipse, forming a thin slice. 'Say the area in this triangle is the same as the area in another one, like this.' He dipped the quill in the inkwell and drew two more lines, this time extending in the opposite direction from the dot to form a wider triangular segment from the Sun's position to the nearer end of the ellipse. He looked up to see if he were being clear.

'Go on,' the older man urged.

'The data shows that both of these areas are swept out in equal times, meaning that the further the planet is from the Sun, the more slowly it goes.'

'Because it has a smaller distance to travel between the two lines,' said Mästlin, pointing to the base of the thinner triangle.

'Precisely. Planets move faster when they are closer to the Sun because the force moving them is stronger.'

'And all this is in your new book?'

'All of it. I have called it *Astronomia Nova* because I believe it is just that: a new astronomy. I am now investigating the differences in the speeds of each planet at closest and furthest approach.'

Mästlin looked stunned. 'Astonishing,' he murmured.

'If I am to complete this work, I need security. My family grows apace, and Prague is growing restless . . .'

Mästlin snapped out of his daze, perhaps guessing where Kepler was heading. 'I heard that Rudolph signed a peace treaty with his brother last year. They divided the Empire between them.'

'They did, but it will not last. Rudolph has been trying to silence Matthias's claims to the throne for years. Last year, Matthias rebelled. He raised an army and marched on Prague but they were stopped outside the city by Rudolph's forces. They carved up the Empire and all seemed quiet for a while, but Rudolph was forced to rely on Protestant armies from the neighbouring estates to counter his brother. In return for their men, the estate owners secured religious freedoms for Lutherans that place the Emperor at odds with Rome. He's still smarting from that embarrassment because the estates fall within the Empire, so he should simply have commanded their loyalty, not had to bargain for it. There are now rumours that Rudolph wants revenge on the estates to restore his standing with Rome and look strong next to his brother. I believe the rumours are true. So, I am looking to move my family, and I would like to move this way.'

There, he had said it. He watched Mästlin carefully, hoping for some flicker in his eyes that might indicate a favourable reaction, but his former tutor was looking at the desktop.

'There are those here who remember your theological debates. The way you disagreed with some key Lutheran principles.'

'Magister, I have made no secret of the fact that I cannot believe in the transubstantiation. The metamorphosis of bread and wine into Christ's body and blood is too much like magic.'

'Which is why we have the ubiquity doctrine: that God is interwoven into the fabric of all things and that is how we receive the blood and the body of Christ through the sacrament.'

'You know that I cannot believe in that either. It is still too close to magic for my liking. The bread and the wine are symbolic. We receive Christ's special help at the time of communion through our prayers.' Kepler slumped back in his seat. Nothing had changed here in the fifteen years he had been away. They were still as intransigent as ever, still missing the whole point of Lutherism.

'Europe is heading into darkness,' said Mästlin, standing up. 'This is no time to hold an ambiguous position.'

'My position has been clear since I was a student.' Kepler also rose to his feet.

'Your position is ambiguous because you disagree with the ubiquity doctrine – one of our foundation stones – and worse, you take your stance on it from the Calvinists.'

'In this one thing alone, yes, I agree with the Calvinists – but on nothing else.'

'I wish you well, Johannes, but I fear that there is no place for you here, not until we can count on your naked faith in Lutherism.'

'But I need support if my ideas are to spread. I plan to send my book to academics and princes across Europe but I don't have many copies. I had to use much of the money Rudolph gave me to publish the book just to buy food. You could at least teach my theory of elliptical orbits or include it in the next edition of your *Epitome Astronomiae*. Let the truth be known.'

Suddenly Mästlin looked angry. 'I will continue to teach Ptolemy's methods.'

'I can scarcely believe my ears; it was you who taught me Copernicus in the first place.'

'In private session, Johannes, not in public. Times have changed, hardened. Your name . . .'

'But this one idea is a bigger monument to God than any cathedral.'

Mästlin jabbed a finger at Kepler. 'You name is tainted. You are not trusted any more.'

An iciness lodged itself inside Kepler. He searched Mästlin's face for some hint of compassion. 'I take it that's why you have not replied to my letters these past five years?'

His old tutor looked away.

'Magister, I can only follow what I believe to be true – in astronomy and in my faith. One feeds my mind, the other my spirit. Neither can be taken in narrow slices.'

Mästlin pulled open the door. 'And that belief is precisely why you must leave. Obedience to the Church is all that matters now. Return to Prague and never come back here.'

As the memory played itself out, Kepler looked at Barbara. 'What else can I do Barbara? Tell me!' His voice was sharp.

'Maybe it would help if your book didn't have a foreword saying that none of it was true.'

Kepler flinched. She could polish words into daggers. Her sharpest were always the truest.

After discovering that Tengnagel had done nothing with the ledgers, Kepler had opened negotiations through the Imperial Court to have them returned to his house, arguing that he could then complete the Emperor's bidding and produce *The Rudolphine Tables*. The ledgers reappeared at a particularly inopportune moment, one not dissimilar from tonight, with Barbara brooding in her chair and the children packed off to bed. The bow-legged deliveryman had heaved pile after pile of books from the back of his wagon into Kepler's hallway, wedging open the door and letting in the chill.

'Won't you even carry them to my study?' asked Kepler. 'It's just down the passage.'

The man's accent was rough. 'I've my instructions from Lord Tengnagel. And more jobs to do after this one.'

So Kepler hauled them out of the doorway and along to the study. Over the course of the following week, he moved them upstairs to the spare room, where there was more space. He then neglected the tables in favour of mining the data for his own ends.

Tengnagel had agreed to return the ledgers on one condition: that Kepler would seek his approval for everything he wanted to print using the data. At the time, Kepler would have agreed to anything just to have the precious observations back, but his attention had once again been diverted from the *Tables* by the lure of Mars. When Tengnagel read *Astronomia Nova* and realised that Kepler was not advocating Tycho's planetary arrangement, the Junker had insisted on writing a foreword clearly stating that Kepler had misused the data to suit Copernican ends.

The poisoned words had been handed to Kepler, the ink still wet, moments before he was due to set off with his manuscript to Frankfurt. Without the damning foreword, he would be denied the right to publish the book at all. What could he do?

He had hoped a publisher would share the cost of production, allowing him to print all the copies he wanted, but, as he went from table to table in the dusty hall, his mood became increasingly desperate. The first words the publishers saw in the manuscript told them to ignore the contents. As they read that, their eyes would glance over the top of the manuscript, checking to see if this was a joke. Kepler had assured them, pleaded with them, it was not. He had toured the taverns dispensing ale and bonhomie with them but to no avail. Times were hard, he was told. Yet not so hard that a few key manuscripts by other authors were auctioned off, he noticed.

So, he moved on to Heidelberg, where the printing began using the residue of the Emperor's money. He promised the ink-stained printer that more funds were on the way. Kepler decided that the best course of action was to disregard the ignominy and to send copies to those most likely to appreciate his work, such as astronomers and mathematicians, and also to those most likely to return favour, such as kings and dukes of distant lands. He just prayed that something would arrive soon to help him meet the debt.

He had done his best to save the situation, so why did Barbara keep bringing this up? Did she not know how much it embarrassed him? His anger simmered.

'I could earn a king's fortune and you'd still not be happy. We are hardly peasants, but somehow you fail to be able to understand this.'

She resumed staring into the empty grate. A single tear rolled down her cheek as, outside, another coach rumbled closer. Kepler paid the vehicle no more than a moment's attention until it stopped right outside their house. In the moment of stillness that followed, a horse snorted and jangled its harness.

Barbara looked up. 'Not again. What does His Precious Majesty want now?'

There was a rap on the door.

Kepler got up from his chair when the knocking came again, louder and more urgent.

'Quick,' Barbara snapped. 'Before they wake the children.'

Kepler hurried to the front door. It was dark, and he fumbled at the latch. A blast of cold air, and there was Von Wackenfels waiting in the porch. 'Four new stars have been discovered,' he said, 'Four!'

'What?'

'Four new stars.' He waved his hands upwards.

There were a few stars visible in the gap of sky above them but not enough for Kepler to get his bearings. He stepped out, taking a few paces before the cold and damp registered in his brain to tell him that he was not wearing shoes. He dashed back to the hallway and retrieved his leather slippers. Then he ran for the riverbank, one eye on the road, the other on the sky, looking for any glimpse of the new celestial objects.

Von Wackenfels trailed behind him, panting. 'Word reached court this afternoon. I came as soon as I could get away. The stars have been found by an astronomer in Padua called Galileo.'

'Galileo! I once exchanged letters with him, many years ago now. I was planning to send him a copy of my latest book.'

'He says the stars are in the vicinity of Jupiter.'

Kepler reached the riverbank. From the chill wind coming down the valley, it was clear that the air might frost again before spring could truly lay claim to the land. Kepler looked up, running his eyes across the sky to Jupiter. There was a building in the way, possibly the university.

'We need to be on the bridge.'

Von Wackenfels had only just arrived at his side. 'Johannes, why are we heading for the bridge?'

'To see the new stars, of course.'

'No, you misunderstand.'

'Misunderstand what?' Kepler stopped running.

The courtier bent forward, hands on knees, and started talking in between large gasps of air. 'You didn't let me finish. You cannot see them with the naked eye. Galileo has used some kind of looking-glass.'

'What? Like spectacles?'

'A long tube with lenses inside, judging by the description that the Tuscan ambassador brought with him.'

'And this extends human vision?'

'So it would seem.'

Kepler returned to von Wackenfels. 'Where can I get one? Is the ambassador still awake? Can he take a letter to Galileo?'

'Calm, Johannes. I will tell you everything I know.'

They returned to Karlova Street. The carriage was still outside the house, its driver hunched in a blanket and the horse dozing.

'May we use your carriage for the discussion? My children are in bed and Barbara doesn't like noise in the house at this hour.'

Von Wackenfels opened the carriage door. 'Be my guest.'

The inside was no warmer than outside but at least it sheltered them from the wind. Kepler shivered. 'Stars that cannot be seen with the naked eye. Who would have thought such wonders were possible?'

'The ambassador says that Galileo has made further discoveries.'

Sobriety crashed back into Kepler. 'This is a jest. You're making a fool of me.'

Von Wackenfels gave an urgent shake of his head. 'Galileo has written a book.'

Kepler cocked his head. 'A book? That I would like to see.'

A few days later, a leather-bound volume arrived at court. Von Wackenfels sent for Kepler and ushered him into a small office where the slim book sat on the table. *Siderius Nuncius*, the Starry Messenger. Kepler picked it up at once and ran his fingers over the title page. His heart beat faster.

Von Wackenfels placed a hand over the front cover, preventing him from opening it further. 'Before you jump in, you should know this. The Emperor is hearing doubts about the reliability of Galileo's observations. Some are saying that this is all just an elaborate illusion.'

'But we have seen new things appear in the heavens before.'

'Yes, we have, but these are powerful voices, very persuasive. I tell you because what you advise will form the basis of His Majesty's position on these claims.' Von Wackenfels held Kepler's gaze. 'We cannot let our own desire for novelty colour this judgement.'

'I understand. I'll be cautious.'

When he was alone, Kepler placed the book on the table and sat down. He tapped his fingers on the cover, afraid now to open it. Then, decision made, he turned to the first page and began to read.

When von Wackenfels returned, Kepler could not tell how much time had passed. He found it difficult to speak.

'Well?' asked the Privy Counsellor.

'I have tried to be sceptical but . . .'

'Yes . . .'

'Everything has changed.'

'He is trustworthy?'

'He writes with the pen of truth. There is no façade in this work, I can tell. The four new stars associated with Jupiter – he draws their position night after night and shows clearly that they are in orbit around that mighty orb. They are moons of Jupiter. Perhaps there are moons around the other worlds too, waiting to be discovered. And there is so much more he has seen. The Milky Way is a river of stars stretching through space and, perhaps most importantly of all, our own Moon has mountains . . .'

'Some say his claims are outrageous.'

'The wonders in this book are beyond anything a human could imagine on a whim. If there was any doubt before, this proves it: Aristotle's universe is smashed into a thousand pieces.'

Von Wackenfels held up his hands in surrender. 'You forget, my friend, I am no philosopher. What is Aristotle's universe?'

'In the fourth century BC, Aristotle described the universe as being made of four base elements: earth, water, air and fire. Earth necessarily sits at the centre of everything because it is composed of the heaviest element. Water sits on earth; air on water; and fire rises above them all. Then Aristotle introduced a fifth element, the ether. This divine substance floats above everything else and makes the heavens. It is condensations in the ether that we see as the celestial objects.'

'I still don't see what you're getting at.'

'If there are mountains on the Moon, it cannot be made of the ether, which Aristotle said was a perfect substance. The Moon must be made of earth, perhaps water, air and fire, too. If it's made of all these heavy elements then, according to Aristotle, it should be drawn to the centre of the universe. So, why has it not come crashing down on us? Don't you see? The fact that the Moon stays up there proves that Aristotle cannot be right. Everything we thought we knew about the celestial bodies is wrong.'

Von Wackenfels chewed his bottom lip. 'You cannot be mistaken about this, Johannes. As yet, there are no other astronomers prepared to back Galileo.'

'Very well, I will write to him and ask him directly for a list of people who can confirm his discoveries. Of course, if I had one of the optical tubes, I could see for myself.'

Von Wackenfels rolled his eyes. 'I should have anticipated that request. Very well, I'll see what I can do.'

'It is Johannes Kepler, is it not?'

The stranger's voice took the astronomer by surprise. He had ducked into the tavern on the way home in the hope of avoiding the nightly battle of sending his children to bed. Also, he wanted to savour the thrill of Galileo's discoveries in peace.

Insulated from the outside world by the noise of the other patrons, Kepler sipped his ale and contemplated what he had read in the *Siderius Nuncius*. As he did so, a new thought seeped into his mind: had he been duped? What if it were indeed a colossal prank? Why would God create celestial objects and then hide them away from man's sight for so long? What if Galileo was on a par with some of the more obvious flim-flam artists who arrived at court claiming to be able to create miracles?

In the market square Jerome Scoto huddled inside a wooden booth. Once he had been the greatest alchemist in Prague, advising Rudolph and directing experiments in the fume-ridden basement of the castle. Now, after making the mistake of being caught in some trick, he was reduced to peddling crystals to the unwary for loose change, the coloured beads supposedly imbued with mystical power.

Kepler knew that he must urgently write to Galileo requesting confirmation. He could not afford to suffer the same fate as Scoto.

'It is you, isn't it? What a stroke of luck,' the voice said.

Kepler glanced up. The man looked respectable enough, dressed in a black suit. Twin beads dangled from the ends of his collar, keeping them flat against his chest. He was young, with an honest face and a clear complexion.

'Do I know you?' asked Kepler.

'No, sir, you do not, but I have been sent to find you with an important request. I represent a powerful lord, who would have you cast his horoscope.'

'I'm rather too busy to play the prophet at the moment,' said Kepler, thankful that this conversation had not taken place at home where Barbara could have heard him turn down work and thereby money.

'You misunderstand me, sir. My patron is a most extraordinary man.'

'And you misunderstand me. My work is intended for people who understand philosophy, not for those infected by credulity, who think that an astronomer can pluck eventualities from of the heavens. The stars wield an influence over us certainly but not a directive kind, more suggestive. Only a man's true resolve can determine his actions. Only an intelligent man can benefit from the kind of true astrology that I practise.'

'I think my patron is such a man as you describe.'

'Who is he?'

'Ah! The one thing I may not reveal to you. My master wishes to remain anonymous for . . . personal reasons.'

'Is he an enemy of the Emperor?'

'He is most noble, sir.' Kepler noted the response did not actually answer the question.

'This would be the most private of transactions,' pressed the gentleman.

'How old is he? I have to know that.'

'Twenty-five.'

Perhaps too young to be scheming, thought Kepler.

'He would of course pay you well for such expert advice.' The man produced a purse of money and rested it on the table. It looked full.

Vulgar, but effective.

Kepler indicated an empty chair. 'I'll need his precise date and time of birth.'

He made arrangements to hand over the chart a week from that meeting, saying that he did not want to rush things for such a noble client. In truth, he drew it up that very evening, intending to go to bed and perform the interpretation in the morning. Yet when he saw the arrangement of planets, something about the chart made him pause.

Then he realised, he had seen it before; if not the exact chart, then at least one very similar. He pulled down from the shelf his file of reference charts and blew the dust from the top of the pages. Flipping through them, he stopped at the one he had remembered. It was indeed similar but not identical, which was a relief. This was the chart he had drawn up, just for practice, relating to the previous English monarch, Queen Elizabeth.

Both this and the new chart displayed the planets suspended above the horizon on either side of the sky, caught either rising or setting. The aspect was perfect for balancing the individual, preparing them for greatness.

Excited at the prospects for his client's future, he wrote his conclusions in a gush. He then spent the rest of the night trying to deduce who the mysterious nobleman might be.

17
Padua, Republic of Venice

Lights were still burning inside the small house when Galileo arrived. *Just like Marina, a creature of the night.* He paused in the porch, wondering whether Giuseppe would be with her. Deciding it did not matter if he were, he knocked. From force of habit, he knocked gently – the way he used to.

It was some moments before the door opened a crack. 'Galileo.' Marina's voice was soft, as if she were trying not to wake someone. She wore a simple silk robe, tied at the waist, that clung to the curves of her body. Galileo could see she wore nothing underneath.

'I need to talk to you,' he said.

She opened the door. 'I knew it would be you.'

The top of her head barely reached his shoulder. Her chestnut hair fell in waves to halfway down her back and her dark eyes were as potent as the day Galileo had first seen her, crossing that bridge in Venice a dozen years ago. Motherhood had rounded but not damaged her figure.

He walked past her, into the house. In the main room, embers glowed orange in the fireplace, the last of their warmth all but lost to the room. He stopped at the table and helped himself to a brandy, drank it straight down and poured another.

'Bad night?' She placed herself carefully on the upholstered settle.

'Fools, the lot of them, too full of wine to see straight. Most of them are convinced I put crystals in the tubes to trick them into seeing stars. Old Libri didn't even come to the gathering. Libri! Fool. The rest of them spent the whole time complaining of the cold.'

'Perhaps they're not as well equipped as you to fight the chill,' she said passing her gaze over his paunch.

'Is a man to be ridiculed for being successful enough to eat? I tell you, Marina, I'm learning as much about human stupidity as I am about the heavens. How could they not see the new stars? If only the night had been a little clearer. Oh, listen to me: if only this, if only that.' Galileo rubbed his temple, letting his fingers roam upwards to touch his receding hairline. 'So many now speak out against me that Grand Duke Cosimo's reputation is being damaged by his support for me. Even Kepler – Rudolph's imperial mathematician – is writing asking for witnesses to my discoveries, telling me of his need for an optical tube.'

'Then send him one, let him verify your claims and put an end to all this fighting.'

Galileo scowled. 'It's not that simple. I can't risk another astronomer owning an optical tube yet. Not before I complete my survey. These discoveries are mine to make.'

'But you've sent so many out.'

'To dukes and nobles, men of no initiative who will look at what I tell them to look at. I dare not send one to another astronomer, especially a Protestant astronomer. Yet, I can't ignore him. He's the only one who's shown any inclination to believe me.' A thought struck Galileo. 'Perhaps I could send Kepler details of my latest discoveries coded into anagrams. That way no one can attack my claims, but should someone else stumble over them while waving around an optical tube, Kepler can prove my priority.' Galileo grinned. 'And I hear that he wishes to leave Prague. I could recommend him to fill my post here at Padua.'

Marina tensed at the mention of Galileo's impending departure. He cursed himself for stumbling into the subject so inelegantly. 'That's really what I came here to talk to you about,' he said. 'I've secured places for the girls. They'll be safe in the convent until I can find them suitors.' He knew he should tell her the whole truth, but the words failed him.

She nodded, hugging herself. 'When do you move to Florence?'

'The Grand Duke's litter arrives for me on Thursday. I'll spend some time settling in and then I am to visit Rome, to present the optical tube to the Jesuits, maybe even to the Pope himself.'

'How often will you return to the university here?'

'Hardly ever.'

Marina looked so vulnerable. He lowered himself next to her, feeling the comfortable familiarity of her presence engulf him.

She seemed to feel it too. 'Why did we never marry?' she asked.

'There was no need. Our relationship was never a secret. I acknowledged the children.'

'But you never let us live with you.'

'I never saw another woman in all that time. I loved you all too much for that.'

'Oh, I dare say you've been a better parent to Virginia than I, instructing her as if she were a son, teaching her music . . .'

'There was nothing more to be gained by marriage.'

'Especially to a girl with a Venetian accent.'

'That's unfair.'

Her expression grew distant. 'I used to wait with butterflies in my stomach for the Friday boat from Pisa. Every week I was convinced you wouldn't be there waiting for me at the other dock.'

'And every weekend I was, until you fell pregnant with Virginia and I moved you here to Padua, so I could be close to you both.'

She glanced around the room. 'And here I still am, as close as I ever got to living under the same roof – even after Livia and then Vincenzio were born.'

'How is my boy?'

'He's well. Giuseppe is good to him.'

'I've made arrangements to continue transferring money to you for his upkeep.'

She nodded without looking at him.

'Marina, there's something else I must tell you. It's about the girls' convent. It's not here in Padua. It's in San Matteo . . . in Florence.'

'Florence!' Her head swung round.

Galileo got to his feet and leaned against the mantelpiece. 'You can visit,' he said over his shoulder.

'Promise me you will find them husbands, Galileo. They mustn't become nuns, Livia particularly . . .'

'The Grand Duke's court is the best place for me to search.'

'. . . she has too much spirit. You've always said how alike we are. I couldn't become a nun.'

'So I remember,' said Galileo, calling up a hint of her secret smile and a cascade of intimate memories. The ache inside him blossomed into longing. If only he had not become so busy. His neglect of her and their family was never because he did not care – and for years he had managed to see them at least once a week. 'Marina . . .' He caught himself just in time, before he said something he would regret. He could so easily fall in love with her again. What did he mean *again*? He had never fallen out of love, just become ever more distracted by his work. Now his reward was a place at the court of Grand Duke Cosimo II in Florence, and with the optical tube he had a real chance of greatness. If he did not take this opportunity, he would hate himself, or, worse, resent Marina. Yet there was no way he could take her with him. She was not used to such circles and grandeur. She had simple needs and would flounder, and that would reflect badly on him. He swirled the last of the brandy and tipped it into his mouth.

The moment hung between them.

'I didn't set out to find a replacement for you,' she said, 'but you had grown so distant.'

He swallowed the apology that sprang to his lips. 'I could see that you were attracted to Giuseppe the day he brought those first lenses.' The truth was that Galileo had quietly encouraged the affair, by arranging for the lens grinder to make his deliveries to Marina's house rather than to the university where Galileo worked. As the work to develop the optical tube had dragged on, so Marina's distraction had provided him with time. 'I don't blame you,' Galileo said. 'You're the same age as each other, and I grow older every day.'

Galileo avoided polished surfaces now, and had done so ever since the morning he had caught his reflection to see his father's ghost returned to haunt him. The snow-white streaks in his dark beard and the way his skin had toughened were all too familiar. His dark eyes remained the same but were now set below a thickening brow and tangled eyebrows, also reminiscent of his father's.

'I miss you,' said Marina.

I miss you too, he wanted to say. 'I ought to go.'

'Of course,' she said without conviction, her eyes wide and inviting.

They had not slept together since Marina invited Giuseppe to her bed. Surprisingly at the time, Galileo had not missed the intimacy but, for some reason tonight, the lost passion was almost unbearable.

Marina slowly opened the front door, then reached up to straighten his hair. Her fingertips sparked against his skin. 'Age does you no harm, you know. You're still as handsome as when I first knew you, maybe more so. Distinguished.'

He leaned down and quickly planted a kiss on her forehead. 'Goodbye, Marina.'

18
Rome, Papal States

All that remained of the midday meal was the smell of garlic hanging in the air. Father Grienberger threaded his way through the early afternoon crowds, using his height to see over the bobbing heads of those returning to work. Under his arm he carried a copy of the *Astronomia Nova*.

He had been feeling heady ever since receiving the book from the Jesuits in Prague, where there had been copies floating around the Imperial Palace. His excitement had quickened when he realised Kepler's epiphany: ellipses instead of circles. He found himself returning to the book time and time again until he had fixed the new meaning in his head. It was such a simple solution to such a complicated problem. All previous scholars had been hidebound by the unfounded assumption of the perfect circular orbit, the shape given credence only by its antiquity.

In addition, there was the news of Galileo's discoveries, which had run through the college like a forest fire. Times were changing, Grienberger could feel it, but to assimilate this new knowledge was going to be difficult. He needed advice.

There he is!

Bellarmine was sitting at a small table on a terrace, spooning food from a bowl. His red vestments set him apart from the merchants in their doublets, concluding their lunches over one last glass of wine.

Bellarmine raised the spoon again, a look of serene pleasure filling his face as he savoured the mouthful, and Grienberger thought how silly the cardinal looked relishing food in public. He approached, casting a shadow across the table.

'Father Grienberger,' said Bellarmine with surprise.

'Cardinal Pippe said I would find you here. I wanted to . . . Is that ice cream?'

'One of the greatest inventions to have taken place in my lifetime. Sit down and enjoy some with me.'

Somehow the thought of the great theologian slipping out to indulge himself with ice cream was faintly disturbing. 'No, thank you.'

'Do at least sit so that I can stop looking up into the sun at you.'

Grienberger perched, feeling uncomfortable now at being in such a public place and knowing what he had to say. He realised that in his excitement to find Bellarmine, he had not planned how to broach the subject. After a moment he said, 'Remember when Father Clavius asked you to think about reconciling the new astronomy with theology. Well, the time has come.'

Bellarmine put down his spoon. 'Explain.' After listening to Grienberger's animated explanation, he leaned back. 'So Kepler stumbles on a mathematical trick to get the right answer. He *assumes* the Earth moves but cannot find any evidence to *prove* it. I think we need not concern ourselves,' he said easily.

'You don't understand.' Grienberger tried again to explain the simplicity of the system, the accuracy of its predictions, the obviousness of it all from a mathematical standpoint, and its reliance on the Sun being the centre of everything. Yet, when Bellarmine's expression darkened, he stumbled over his words and eventually became tongue-tied. He riffled through the book, placed it in front of Bellarmine, and pointed out a passage.

> *Piety prevents many people from agreeing with Copernicus out of fear that the Holy Ghost speaking in Scripture will be branded as a liar if we say that the Earth moves and the Sun stands still. But these persons should bear in mind that we learn the most important things with the sense of sight, and therefore cannot detach our speech from the visual sense. Thus, very many things happen every day when we talk the language of the sense of sight even though we know for a certainty that the situation is otherwise.*

Bellarmine scowled. 'What is this gibberish?'

'There's more.' Grienberger thumbed to another passage.

To teach mankind about nature is not the purpose of Holy Scripture, which speaks to people about these matters in a human way in order to be understood by them and uses popular concepts. Why is it surprising then, that Scripture also talks the language of human senses in situations where the reality of things differs from the perception?

'I have heard enough.' Bellarmine angrily slapped the pages over until the book was shut. 'You admire this? Does Father Clavius believe this work?'

Grienberger nodded gravely. 'The mathematics of it, yes. The theological justification is down to you.'

'It is heresy. The Holy Spirit did not lie when dictating the Bible. Tell me, what do you know about an astronomer called Galileo? The Inquisition is hearing disturbing things about him. He's sending "glass tubes" to nobles throughout the Italian peninsula and beyond, currying favour by claiming to see new planets. He's apparently putting other discoveries in coded messages. Yet, he's sending nothing to Rome. What are we to make of this behaviour?'

A serving maid approached the table, intent on clearing away Bellarmine's bowl. He made to stop her but realised that his ice cream had melted. He slid the bowl towards her and his look of displeasure intensified.

'Can we even believe his discoveries? I have heard that Father Scheiner in Ingolstadt is saying that the results are nothing but trickery, that he has seen through one of these but can see nothing.'

Grienberger chided himself for being so hasty in seeking Bellarmine's help. He had hoped to keep the next piece of information secret for a little longer, at least until the Jesuit craftsmen had succeeded in their work, but he could see that there was only one thing that would satisfy Bellarmine now. 'Come with me, Cardinal.'

In Grienberger's office was a long thick tube that had been carefully split lengthwise and opened out on his desk; inside it were a pair of curved lenses, one at either end of the tube.

'This is all that one of Galileo's optical tubes is?' asked Bellarmine.

Grienberger nodded. 'We intercepted it en route to the Duke of Bavaria.' He picked up the lenses and handed them one by one to Bellarmine, who lifted them to his eye and waved them around trying to get them to focus.

'Pure glass, no tricks,' said Bellarmine. 'Have you seen the things that Galileo describes?'

'From the roof of the college before we took the tube to pieces. They are real. We're now making our own optical tubes. We'll rebuild this one and send it on to the Duke. We have arranged for it to be loaned to Johannes Kepler.'

Bellarmine looked warily at Grienberger, who explained. 'Kepler has been asking Galileo to provide him with the instrument, but Galileo has refused, making the excuse that he needs them for princes. It's as if he has something to hide. Father Guldin has agreed to make the arrangements in Prague. He was a Lutheran before converting to us so Kepler might view him sympathetically.'

Bellarmine replaced the lenses. 'You should think yourself lucky that Cardinal Pippe is not here: doing favours for Lutheran heretics . . .'

'We still hope to attract him to our ranks. From what I hear, Prague grows more unstable by the day.'

'Indeed, but would we want Kepler any more? We may end up converting him just to tie him to a stake in the market square.' Bellarmine rubbed his baggy eyes. 'Father Grienberger, I must know what we are dealing with. I trust the evidence of my eyes in preference to the pen-strokes of a Lutheran, no matter how eminent you say he is. As for Galileo . . .'

Grienberger spoke quickly. 'We have already invited him to Rome. We're as anxious to meet him and understand his agenda as you.'

'When were you going to tell the Inquisition this? Oh, no matter, when does he arrive?'

'We expect him sometime in the next few weeks.'

'Good. Let us see what Kepler makes of Galileo's findings – and let us see for ourselves what Galileo has to say. But tread carefully, Father Grienberger, you're on dangerous ground.'

19
Prague, Bohemia
1610

The splintered wood pulled at the soles of Kepler's stockinged feet as he tiptoed up the stairs. In one hand, a feeble candlelight writhed in the draught from the attic rooms. In the other, he held a wooden tripod, and a most precious leather tube containing a pair of ground glass lenses.

It was the Duke of Bavaria's optical tube; he had arrived at court carrying the instrument, and for as long as the anxious political negotiations continued, Kepler could make use of it. The Duke's Chancellor, Hewart von Hohenburg, had brought the tube to him, carefully explaining that a Father Guldin had pointed out Kepler's need.

Father Guldin. The Jesuits again.

They had asked for nothing in return, though they must know politeness alone would compel him to send them his observations. As much as he wanted to believe that they were motivated solely by their brotherhood in astronomy, since Grienberger's attempt to convert him in the wake of Susanna's birth he could not shake his unease at their patronage. Their fears for Susanna had been entirely unjustified, he thought with some pride. She was strong and clever, with brothers who were equally robust.

He had left his shoes at the bottom of the stairs in the hope of being quiet. Frau Bezold's footsteps often woke him when she rose early to begin her chores, and the last thing he wanted was to disturb Barbara. For once, she had quietly dropped off to sleep and was now snoring softly in the marital bed.

At the top of the landing, he used his elbow to slip the latch on the back room and crept inside. As he did so, the room filled with movement and an awful wailing. Kepler felt the breath rush from his lungs as the spectre rose into the view. It was a seething mass of bedsheets.

As suddenly as the terrible noise had started, so it stopped, and Kepler found himself staring at a familiar face, old and pinched, topped with a skullcap of muslin that almost hid the wiry crown of grey hair. 'What are you doing here?' the face demanded.

'I prefer to ask you that question, Frau Bezold. This is not your room.'

'It's quieter in here at the back. I can sleep better.' She pulled the blanket tighter around her shoulders. As she did so, a small wooden crucifix escaped the fabric of her nightwear.

Kepler stared, shocked by the Catholic symbol. The housemaid tugged the blanket harder to cover the necklace, as if to remove it from sight would erase significance.

'Frau Bezold . . .' He set down the candle. It flared briefly and then settled to a steady flame again.

'It's the only way since Gerhard died. I like to talk to him on Sundays, but when they stopped us going to church . . . How else could I . . . Don't you miss the sacrament?'

'I've learned to live without it.'

'Well, I can't. It's all the same God, isn't it? They make it easy for you to change, you know. Just go to the church and answer a few questions. It's not as different as you think.'

Kepler sighed. 'You must follow your conscience. But you are wrong to think that the two Churches are in any way similar.'

She sniffed loudly, readjusting the blanket again. 'So, what are you doing in here?'

He almost laughed at her defiance. 'I have the new instrument from Italy.' He rolled the red leather tube in his hands, admiring its embossed feathers. It felt incongruously light for something so important.

'Fancy,' she said. 'What does it do?'

'Makes far away things seem closer. They say it reveals new wonders in the heavens.'

'What good is that? There's nothing wrong with what we've always been taught. Earth lies at the centre with seven heavenly spheres above our heads and seven of the Devil below our feet,' she recited. 'How we behave in this life determines whether we go up or down in the next. Why change things?'

'Because it doesn't work. Here, hold this,' he handed her the instrument, 'and don't drop it.'

He busied himself with the tripod as she lifted the tube to her eye and swept it around the room. 'I can't see a thing.'

He took it back and set it on the tripod, swinging it gingerly to test it was mounted securely. With a grunt of satisfaction, he pulled open the window, admitting gusts of cold air into the room.

Frau Bezold made a *brrrrrr* sound. 'I'm going to bed, it's too cold.'

'Aren't you even slightly curious?' he indicated the leather tube.

'What good will it do me?'

'It will show God's realm to you. How can you not want to see that?'

'When you get to my age, you're glad of every postponement.' She shuffled from the room.

Kepler aimed the tube at the shining orb of Jupiter but when he squatted to look through, there was nothing but blackness. He nudged the tube. Still blackness. Again and again he tried, the movements becoming a little larger, a little more desperate. Occasionally something would dart across his line of vision, but he could not stop the tube in time to catch it.

He stood up, tight across the shoulders, and wrung his hands to warm them up, then realigned the instrument. This time Jupiter sprang out clearly. Kepler stared in amazement as four dimmer points also solidified in front of his eyes. He wanted to laugh and cry simultaneously. The last time he had felt this way was when Barbara gave birth to Ludwig.

As he looked, he noticed that Jupiter appeared bigger than the four new stars, more of a disc than a spot of light. It was said that the new stars moved around Jupiter, creating their own system of revolution, and that if he watched all night he could see them move.

Did they, too, move in ellipses? Could his laws of planetary motion be applied to them? It seemed a reasonable assumption. To prove it, he would need someone to devise a way of measuring the positions with this new device.

He felt a sudden pang. Tycho. What would he have made of all this?

A new thought chased away the old. He hurried to the children's bedroom and woke Friedrich and Susanna, shushing them to silence. 'I have something for you to see,' he whispered, ushering them upstairs as quietly as possible.

Inside the darkened attic room, he placed a chair for Susanna to look through the tube. Friedrich traced his hands along a leg of the tripod.

'Don't move it, son. It must be completely still or your sister will see nothing.'

The boy was a miniature version of his mother, blessed with her apple cheeks and chubby fingers. Kepler could not help but smile whenever Friedrich was near, wrapping his inquisitive nature around whatever was at hand, just as he himself had behaved as a child.

Susanna stood on the chair, holding its back, and studiously squinted into the eyepiece. She reminded Kepler in so many ways of Regina at that age: her earnest behaviour and her growing imitation of adults. She had even taken over the care of Astrid the rag doll, when her big sister left to be married.

Kepler stood poised next to her. 'Be patient, Susanna, it's difficult to see. You must be very still.'

'Papa, I can see them.'

'My turn, my turn, my turn.' Friedrich rocked the chair.

'Careful!' scolded his sister.

Kepler lifted her down and helped Friedrich clamber up. It was obvious his sister's two extra years of growth were an advantage. So Kepler held him by the chest and lifted him up, but the boy dissolved into giggles rather than making a serious attempt at astronomy.

'That tickles.'

'Oh, does it? I thought this was tickling.' Kepler increased the pressure of his fingers into Friedrich's soft body, imitating Frau Bezold kneading bread. His son wriggled in his arms, squealing in delight.

Susanna joined them, adding her voice to the growing cacophony. Together they tumbled to the mattress, lost in their pretend wrestling match.

'Enough!' A sharp voice cut through their play. 'Enough, I say!'

Barbara stood at the door, hair tousled, a smelly tallow candle in one hand; two-year-old Ludwig balanced on her hip. Panels in the side of her nightdress showed where she had adjusted it to accommodate her new bulk, as if she were still pregnant with Ludwig. His little foot dug into her midriff, pushing the soft fat into a foothold.

'We were just playing,' said Kepler.

'*Playing*? At this time of night? I'll never get them back to sleep now.'

'Barbara, this is a historic occasion. The discovery of these new stars is worthy of celebration.'

'Only in your head. It's a fantasy. It means nothing in reality because it changes nothing. Nothing is different for us. Nothing.'

'I should have known better than to expect a simple country girl to understand.' Kepler pushed past her to stamp as heavily as he could down the stairs. 'You think about nothing beyond your own reach. Well, there's more to this world than eating. Our minds must be exercised or we're no better than the beasts.'

'That's right, walk away from us, just like you always do. Go and shut yourself away and daydream.'

The sound of the children sobbing followed him the rest of the way down to his study.

At court the next day, Kepler struggled with a fearsome headache. For an instant, he thought it might be clearing when von Wackenfels bounced up to him flapping a sheet of paper, calling excitedly, 'My friend, I have more good news!'

Kepler shifted the charts he was carrying and took the letter.

'It's from Galileo,' said von Wackenfels. 'His latest discovery, apparently, but we need your help to tell us what it is.'

Momentarily flattered, Kepler scanned the words. Addressed to Rudolph, they were full of supplication and praise, and he felt a tremendous rage against the Italian.

Why is he not writing these things to me? Kepler thought. *Have I not been the staunchest champion of his claims? Have I not sent him a copy of my* Astronomia Nova? *Why does he favour me with so little in return?* Galileo's one and only letter to him had been a brush-off, the words guarded if not wholly evasive.

As Kepler fought to read on, his fingers clenched the paper into creases. Then he arrived at the discovery: *Smaismrmilmepoetaleumibunenugttaurias*. Kepler fought to suppress his anger as he stared at the nonsense word, coded to protect Galileo from further attacks, yet ready to prove his priority.

'Galileo teases us, eh? We need your brain, Johannes, to uncover what it is he is telling us.'

Kepler's first instinct had been to refuse. *Why should Galileo treat them all like puppets?* But just as he was about to thrust the letter back to von Wackenfels, he realised that uncovering the message would be a victory. *Yes! Unmask the discovery and send it back to Galileo. Perhaps then he will treat me with more respect.*

Kepler took the note home and crept into his study, anxious to avoid Barbara after last night, and spent the rest of the day pushing the letters of the anagram around. He tried a few words of Latin, kept the good ones and returned the bad ones to the melting pot and tried again. If only Galileo's handwriting were clearer; it was difficult to tell 'u' from 'v'. He broke the nib on his quill jabbing it into the desk in frustration when one attempted solution ran out of letters at the last minute. He spent the next few minutes refashioning it with a desk knife. He was dimly aware of the household beyond the closed door but refused all food in order to work on undisturbed.

Then, as night fell, his excitement mounted. This time he knew he was close to the answer. In the same way that he could sense when numbers and patterns were about to fall into place, even before his conscious mind could see the solution, so the letters suddenly felt right.

Then it happened.

Salve umbistineum geminatum Martia proles. There was something barbaric about the Latin verse but it was a solution: Hail, burning twin, offspring of Mars. The meaning was clear: Mars has moons as well.

Exhausted, he let the quill drop and made his way to the attic. It was difficult to locate Mars, but he caught it close to the chimneystack of a house on the opposite side of the road. Lining up the optical tube, he managed a long enough look to be certain of one thing: there were no moons dancing attendance to the planet's red disc. No moons. He had not solved Galileo's puzzle.

Disappointed, he drifted though the house, looking for Barbara. He found her squeezed into a chair in the front room, reading her prayer book. He edged towards her, knowing better than to share his troubles. When she did not shuffle her legs away, he settled at her feet.

'The children missed saying good-night to you earlier,' she said softly.

'I'll make it up to them tomorrow. Did they say their prayers?'

'Of course.'

'You are a good mother to them.'

After a moment, she rested her hand on his shoulder.

The crowd that gathered around the makeshift stage in the market square was rowdy, as usual. The leading actor preened in the evening sunlight, wearing a gaudy suit of orange silk that played tricks on Kepler's eyes, and a white hat with enormous plumage. The outfit was as incongruous to his role as a sea captain as it was in keeping with his status in the troupe.

Those closest hurled abuse or encouragement depending upon their mood, and it occurred to Kepler that the spectacle was only one step removed from the bear baiting that took place in some quarters of the city.

Ordinarily he would not be here, but Hewart von Hohenburg had suggested they spend the evening together, starting with the street theatre. Kepler never missed the chance to meet up, having long ago accepted his predilection for the company of those born into a higher station.

'And what brings the Chancellor of Bavaria to Prague on this occasion? Apart from bringing me the optical tube, of course.' Kepler looked at his friend, hoping that his joke would register, but Hewart looked tired and unusually jumpy.

'Business as usual,' said Hewart unconvincingly. 'Tell me, how is Regina liking marriage?'

'She is the happiest I have ever known her. Philip is a good man, but I do miss her so. Pfaffenhofen is far enough away to deter the casual visit,' said Kepler.

'And how is my godson, Friedrich?'

'I swear that he is God's gift to us for our patience. He is the noblest heir a man could wish for. He has grace and humour; it is impossible to be sad with him around.'

'It's quite a full house you have now.'

'Yes, thank goodness my mother returned to Leonberg. Even so, we are full to bursting. The three children keep Barbara and Frau Bezold very busy.'

'And to think you were so worried about Susanna to start with.'

'My pedigree with children was not a good one. Thankfully things are better now. All three are in the best of health. It is Barbara I worry for. The melancholy has attached itself to her so badly that her black days outnumber her good.'

'She will come round. Childbirth does strange things to a woman. Makes me wonder sometimes why God chose them for the task when men are so clearly stronger.'

An actor playing the fool was drawing whoops of delight from the crowd with his faltering English accent. Hewart spoke, his voice all but lost amid the happy noise. 'I confess my weakness for the theatre, but I am unsure whether it is to watch the players or the crowd. It is a welcome distraction at this time.'

'I take it you saw through the looking-glass before giving it to me?'

'No, well, I say no. We tried, but the truth is we couldn't make it work. Didn't know what we were looking for. It just all seemed so . . . black up there.' He shrugged pitifully. 'So, I burn with envy for you at being able to master it. Is it marvellous?'

'It's a revelation but an acquired skill. It's a challenge to find the object in the first place. When you have, the image appears in a round spot, as if you are peering up a chimney and seeing the small patch of light surrounded by a halo of darkness. It takes time for your eyes to comprehend what you're seeing, but persevere and you will be rewarded. I have seen all that Galileo describes, despite my troublesome eyes. There are more stars in Heaven than we can possibly imagine.'

'Are you in contact with Galileo?'

'Not really, he sends infuriating anagrams to Emperor Rudolph that I attempt to decode. We had to beg Galileo to reveal the first solution. I thought he was telling me that Mars has two moons but instead it is that Saturn appears "three-formed". I have managed only to catch glimpses of this new discovery but I believe there are two close moons around the planet. Recently he sent a new riddle to court, *Haec immatura a me iam frustra leguntuory . . .*'

'These immature things I am searching for now in vain,' Hewart translated.

'Precisely, but I have made no headway rearranging it at all.' Kepler shrugged.

At the unsatisfactory end of the play, which included a five-minute death scene for the vision in orange, Hewart invited Kepler to return with him for a drink.

'Let's go to my house on the way and I'll fetch the optical tube,' said Kepler.

Hewart agreed at once. As they walked to the waiting carriage, the Chancellor grew nervous and appeared to be on the verge of saying something. Kepler pretended not to notice, concentrating instead on holding himself erect – he always felt so clumsy next to Hewart.

Making their brief stop at Karlova Street, where Barbara was talking to Frau Bezold about what to serve Friday's dinner guests, the carriage continued over to the New Town. It deposited them at a large well-furnished house, where they were served tawny port.

They climbed to the top of the house and out onto the wooden roof terrace. Below them, the city was swirling with nightlife. Above, the press of Heaven bore down. The cobalt sky was not yet fully dark, but the multicoloured stars twinkled in the evening air. Not for the first time, Kepler marvelled at the way the vista could be transformed into mathematics by the human mind.

'What is it that attracts us to the stars?' Hewart asked.

'We are made in God's image. Our faculties cannot help but be primed for astronomy. The songbird sings because it is in its nature, so it is with the human mind and astronomy. God has given us the gift of curiosity and the mental faculties to read his words in the architecture of the cosmos.'

Hewart swept his gaze across the starry vault. 'Don't you agree that standing beneath the stars is the only time that certain thoughts can be entertained? The only time that certain things can be said?'

Kepler paused in his fiddling with the tripod. 'What's on your mind?'

Again there was that strange look of anticipation and then the self-conscious smile. 'Nothing,' said Hewart, dropping his gaze to the streets.

'What is it?'

Hewart waited, perhaps wrestling with some internal conflict. Then turning to Kepler he spoke in a whisper: 'Rudolph is working towards revenge over the Protestant estates. He negotiated with my Duke to raise an army of Bavarian mercenaries to march on Prague and crush the Protestant Union. But it's gone terribly wrong. Rudolph hasn't provided the money to pay the army, and they're on the rampage. That's why my Duke and I are here, to warn him of the danger if he continues to renege on the deal. The last report I heard, the mercenaries were moving through Upper Austria, burning and pillaging. If they reach Prague, it will be the excuse Matthias needs to send his army into the capital and wrest control from his brother. The only question is what the Lutherans and the other Protestants will do. Will they defend Prague or attack it with the mercenaries?'

Kepler felt the pit of his stomach fall away. 'How soon before the mercenaries arrive?'

'Weeks, but they have cavalry so it may only be a matter of days.' Hewart looked across the city's darkened rooftops. 'Prague is no place for a young family. It's about to become a battleground.'

20
Rome, Papal States
1611

Summer had come early to the Italian peninsula, carpeting the slopes in a million shades of green. Galileo studied the grand patchwork as he bobbed along in Grand Duke Cosimo's litter, carried by two broad-chested servants. It would be all too easy for someone to see only a single wash of colour but to Galileo the landscape resolved itself like faces appearing in a crowd: trees and shrubs; ferns and grasses; bushes and vines.

Wedged next to him was a narrow box, some four feet long. Inside was what everyone wanted to see: the optical tube. He patted his fingers against the wooden case, as an indulgent father might comfort a demanding child.

The litter slipped into the busy streets of Rome as the setting sun transformed the mighty buildings, statues and obelisks into silhouettes. Weaving around the carts and pedestrians, they reached the Tuscan embassy as the staff were lighting the first torches. At sight of the arrival the staff hurriedly finished their preparations and lined up to receive Galileo. The Grand Duke himself had sanctioned the stay.

The runners placed the chair outside the front porch, then peeled the straps from their shoulders and arched their backs. Galileo stepped from the small chamber and immediately despatched one of them with a message to the Roman College, announcing his arrival. The other one hauled down Galileo's trunk of clothing and lugged it round to the back of the building, a servant guiding his way.

As Galileo entered the spacious hallway the housekeeper handed him a message – an invitation for later that week from someone called Federico Cesi.

Whoever Cesi was, he signed himself the Marquis of Monticelli and requested that Galileo, and the optical tube naturally, be guest of honour at a dinner to be given by a body called the Lyncean Academy. According to the brief explanation, this august organisation hungered for members who pursued true knowledge and would be honoured if Galileo deigned to join their number.

Galileo found the invitation sufficiently intriguing to pen a quick reply in the affirmative and hand it back to the housekeeper for despatch.

When the litter bearer returned from the Roman College, he carried a brief reply inviting Galileo to visit Father Clavius, the Head of the College, the very next morning.

'Did they say anything else? Give you some hint . . .'

'No, signor,' said the exhausted man.

As cockerels crowed somewhere on the outskirts of the city, Galileo dressed himself carefully. He slipped into his newest tabard, created from a beige brocade that in a certain light looked golden – and also did a good job of disguising his belly. He remembered to brush his hair and even pulled his beard into some semblance of shape, smiling as he thought of little Virginia sitting on his lap and twirling her fingers in it.

Little Virginia! She's eleven now, practically an adult. He pushed the thought aside. He had to concentrate. The Jesuits were his stepping stone to the Vatican; he had to convince them of his discoveries.

In the bright glare of the sun, it was hard to believe that, only fifty years ago, Rome had been pillaged. Piles of cadavers had lain reeking in the streets while the invaders slaughtered anyone who tried to retrieve a body for burial.

Now Galileo's litter wove through a city echoing with footsteps and conversation, the grumble of carriage wheels and the clop of horses' hooves. Great rectangular buildings stood firm, curves having been largely replaced by the solidity of straight lines and right angles. There was little adornment around the rectangular windows and doors, nothing to spoil the buildings' defiant faces.

As he drew close to the Roman College, Galileo's eagerness gave way to anxiety. He stumbled as he alighted from his vehicle and wondered whether Clavius was watching from one of those enormous windows.

Walking through the magnificent entrance, the lintel some twenty feet above the ground, Galileo could not remember ever feeling so small. Surely this place had been built for gods, not men.

Jesuits in black robes moved through the lobby. Occasionally one would glance his way; mostly they ignored him.

Galileo stared at the Egyptian obelisk in the entrance hall, feeling minuscule by comparison. He wondered what knowledge lay hidden within those ancient symbols.

'Every time I look at it, I am reminded of the task we still have ahead of us,' said a deep voice.

Startled, Galileo turned to find a large man in black vestments, whose eyes and mouth turned down in a way that unnerved Galileo.

'Allow me to introduce myself. I am Christoph Grienberger,' said the man.

'Galileo Galilei, a pleasure to meet you, Father.'

'The Professor of Mathematics, Father Clavius, is waiting.' Grienberger indicated the way, and set off at a lumbering pace that made Galileo feel quite youthful.

'How much do you believe the ancients knew?' asked Galileo as they walked into the heart of the college.

'Sometimes I think that everything we struggle to uncover was known to the Egyptians; that if we could just read the glyphs, our work would be done.'

'In that case, shouldn't we devote all of our academic efforts to deciphering?'

Grienberger inclined his head towards Galileo. 'Would you be content to do that? If there is one thing I have discovered about learned men, it is that they have a stubborn loyalty to their chosen fields.'

'And their own convictions,' said Galileo, thinking of his father. The man's steadfast devotion to music and his insistence that melodies should reflect the instantaneous mood of the lyric rather than follow some overarching design, had led him into bitter argument with the traditionalists. Nevertheless, his madrigals were still being sung.

They turned into a corridor. The herringbone pattern of wooden blocks softened their footsteps, enhancing the sense of reverence.

Grienberger knocked on a door and immediately opened it wide.

Father Clavius's neck had curved forwards with age. A knotted brow topped his square face; thickets of hair peeped from his ears and nostrils. He was sitting in an armchair near the towering window. Sharp sunlight sliced acrosss his pallid flesh, and his yellowing eyes tracked the new arrivals.

Galileo knelt before him. 'Father Clavius, you have done more to dignify mathematics than any man alive.'

Clavius lifted a tremulous hand from the chair arm and turned his palm upwards. Galileo frowned at the ambiguous gesture.

'Please stand up, Galileo, kneeling is unnecessary,' said Grienberger, moving to stand beside the chair.

Galileo got to his feet and stepped back so as not to loom over Clavius.

The old man spoke, the effort of the action obvious. What his voice lacked in power it made up for in the unmistakable intonation of a man used to being obeyed. 'Someone has to champion the mathematical arts; they have been the poor relation of other pursuits for too long.'

'Indeed they have. I have always enjoyed the company of numbers. They bring with them a sense of security. Father, if you would permit me to return this evening, I would enjoy the opportunity of showing you the wonders of Heaven through my optical tube.'

'That will not be necessary,' said Grienberger. 'We have seen all we need to.'

Galileo looked from Grienberger to Clavius. There was a hint of mischief on the professor's face.

'We have made seeing instruments of our own,' explained Grienberger.

'You have seen the Medici stars?'

Grienberger raised an eyebrow. 'The moons of Jupiter, yes, and the stars of the Milky Way and the strange markings on the Moon. The question is how we interpret all of this.'

'It proves Copernicus.'

Clavius made a noise but otherwise remained impassive.

'That is a bold claim. You refer, of course, to the idea that the Sun is the centre of the universe,' said Grienberger.

'I do,' said Galileo. 'None of my observations contradict Copernicus.'

'Neither do they prove it,' said Clavius.

Grienberger inclined his head. 'What do you make of Johannes Kepler's work?'

'I find him tedious. He approaches the observations with a caution bordering on pedantry. Copernicus completed his work on this subject almost seventy years ago. We have no need of a Lutheran champion on this side of the Alps, pushing around some numbers and claiming victory. Copernicus beat him to it; we cannot let that great canon's work be usurped by a Protestant mathematician. His talk of elliptical orbits is ugly. How can anybody think that the planets move on anything but perfect circles? How can God's Heaven be anything but perfect?'

'I can find no errors in his mathematics,' said Grienberger. 'In fact, I would say that the *Astronomia Nova* is one of the greatest works of astronomy ever published. Kepler's elliptical orbits reproduce the appearances of the planets better than any other mathematical model, including that of Copernicus.'

Galileo gasped. 'Is the original cutting of the cloth less important than the final decoration? Kepler's work is nothing but adornment on that of Copernicus. One can do without the details but take away the original pattern and you are left with nothing.' He clasped his hands together. 'We must have the courage to believe the evidence of our own eyes.'

Clavius pushed himself to his feet, shuddering with the effort. Grienberger immediately bent to support him.

'I know you want to shout your discoveries from the rooftops, but we must move carefully,' said Clavius. 'I realise that the orbs of Heaven need rearranging to accommodate what you have seen. But we must be careful to make the correct interpretation. We cannot allow a casual glance through an optical tube and a snap decision to become the preferred route to knowledge. If we do that – and make no mistake about this – the theologians will crush our little hobby of stargazing for ever. Look at me, Galileo. We cannot risk natural philosophy becoming divorced from religion. Do you understand?'

'But truth is truth, why should we submerge such truthful convictions?' Galileo's muscles began to prickle with passion.

Clavius's jaw trembled as he spoke. 'We are inclined to believe you, Galileo. But the Church is not a court of law. It does not rest on evidence alone. Beliefs, personal preferences and political considerations must all be weighed before we can change such a fundamental piece of understanding. We must persuade the theologians to help us, or we will never succeed. You must comply, or you and your discoveries will suffer.'

Galileo nodded dumbly, unsure whether he had just been threatened or appointed a Jesuit confidant.

As the sun began to slide from the sky, Galileo prepared to meet the Lynceans. He checked the instrument to make sure that the lenses had not been dislodged during the long journey from Florence.

His two bearers looked relieved when Galileo granted them the night off. Soon afterwards, a carriage pulled up. The driver hopped from his seat to open the door. As he did so, he bowed. 'Signor Galileo, Prince Cesi awaits.'

'Prince? I thought he was the Marquis of Monticelli.'

'He is, signor, in addition to being the Prince of San Polo and Sant'Angelo, and the Duke of Acquasparta.'

Galileo was glad that the driver was still stooped in a bow and so could not see the surprise spreading across his face. He climbed into the carriage.

The journey took no more than fifteen minutes. When the driver opened the door for Galileo to step down, a slim youth with an oval face and almond eyes was waiting. About his neck he wore a heavy gold chain with a lynx pendant attached. The wild animal was sculpted in mid-stalk, ears upright, staring straight ahead.

'Signor Galileo, you more than honour us with your presence, you enlighten us. We are all most excited that you are here.' His gaze came to rest on the box.

Galileo was leaning against it, as if it were a walking stick. 'The honour is mine, Prince Cesi, but I think you would have been just as pleased with this box alone, yes?'

Cesi clutched at his heart. 'You wound me with the accusation. What greater pleasure can there be to have the man himself here to demonstrate?'

Federico Cesi carried himself with an easy self-assurance, walking with a casual swing of his arms and laughing at trifles. He was all youth and enthusiasm, usually a combination that irritated Galileo.

Nearby stood a familiar figure in black, wearing a silk biretta on his head.

'Father Grienberger, have you been sent to keep an eye on me?' quipped Galileo.

The Jesuit looked awkward, and Galileo wondered if his joke had somehow struck close to the truth.

'Christoph is a good friend of the Academy,' said Cesi. 'I'd say there's no better mathematician in all of Rome. But he hides his light, takes none of the praise.'

'You credit me with too much, Federico,' said Grienberger.

'Come, let us make our way to the banquet.' Cesi indicated the grassy slope.

Galileo hesitated.

'Tonight, we will be eating under the stars in your honour. The tables are set on the hill. We have even set forks as well as knives. Why let standards drop just because the setting is unusual?' Cesi's smile reached all the way across his face. 'Then lead on.'

The gathering was larger than he expected. Some thirty gentlemen, all drinking and laughing, stood around a wooden pergola. When they were still some ten paces away, a familiar voice rang out. 'I swear there is a new spring in your step, Galileo.'

'What do you expect? I have found new pages in the book of nature. It is enough to reinvigorate any man. If only it could remove the white from my beard, or put the hair back on my head.'

'I think you two must know each other,' said Cesi with another smile, gesturing towards the short gaunt man with heavy black eyebrows who had spoken.

'Indeed, we do. Giovanni Magini, what is it like to finally have been overtaken as Europe's greatest astronomer?'

'I'll let you know when it happens.'

Galileo forced himself to laugh. Magini was ten years his senior and had been chosen over him for the Chair of Mathematics in Bologna. Even though the appointment had been made twenty years ago, the rejection still hurt.

'Allow me to introduce our other guests,' said Cesi.

The circle of introductions rapidly became a blur. One bright young man was an aspiring astronomer. An older man was a philosopher, though Galileo had not heard of him before. And there was at least one doctor among them; he may have been the fat one with the bad skin. The others were various friends of his host, mostly merchants and the odd petty noble with an interest in natural philosophy. No one Galileo took too seriously.

'Looks as though the feast is ready,' said Cesi, pointing to the well-stocked tables beneath the lantern-clad pergola. Galileo could only imagine the effort of carrying the huge wooden furniture up the hill.

The dinner guests ate standing up, something Galileo detested. The ham was good though, and he returned several times for more. He made a small effort with the fork but spent most of his time hoisting cuts into his mouth with his fingers. 'Using a fork takes so much of the pleasure out of eating,' he heard someone grumble behind him. Soon, only Cesi and a few of the younger gentlemen were persevering with the ludicrous implement.

Studiously avoiding the salads, Galileo was cutting another hunk of meat when one of the guests asked him, 'How does it feel to have invented such an instrument?'

Galileo brushed his tongue around his teeth. 'I did not invent the optical tube. I reinvented it.' He relished the puzzled faces before him. The group clustered around him.

'I heard of a Dutch spectacle-maker who had made a looking-glass by placing a concave and convex lens some distance apart. So, I began buying lenses to see if I could replicate the device. It took me three hundred lenses before I found a pair that worked.'

'What magnification do you achieve?' asked Cesi.

'About twenty times, and I think I can achieve more. My experiments are not yet finished.'

'And what made you decide to look at the heavens?' asked the aspiring astronomer, whose name Galileo had already forgotten.

'I'm an astronomer. I look at the heavens. Would you ask a book-keeper what makes him count money?'

The young man looked down at his plate as a ripple of embarrassed laughter circulated.

Galileo lay down his food and swung the carrying box onto a table. He slid it between the half-finished trays of roasted dove and olive polenta, and twisted the catches.

The optical tube rested in a cushion of green velvet. It was just over three feet in length but only a couple of inches in diameter. With its sinuous ridges and fawn-brown colouring, the leather fixed to the tube made it look strangely organic.

Galileo lifted it and held it out, as if presenting it as a gift.

'Is that it?' asked someone in the crowd.

'It does not look very big to see all that way,' said another.

'It is bigger than your eye, that is all that matters,' said Galileo.

'How did this Dutch spectacle-maker conceive of such a device? He strikes me as a very clever man indeed,' said Cesi.

Galileo dismissed the remark. 'I think it must have been an accident. Children playing in his shop, holding lenses to look through. Some folly that in this instance paid off.'

'But the lenses are further apart than a child's arm span. That cannot be true,' said Grienberger, pointing to the tube.

Galileo forced himself to laugh. 'I wait to hear the Dutchman's account of the story, then we can all be satisfied.'

Cesi glanced at the sky. 'It is getting dark, I think.'

Galileo took the tripod from the case and stood it on the grass beyond the terrace. He fixed the optical tube in place and looked up. The night was not as still as he would have liked, but he had observed under worse conditions. Jupiter was shining brightly. It took him only a few moments to capture it in the eyepiece. The men drifted from the pergola to stand around him again.

'Gentlemen, the only people who do not believe me are those who have not looked through the instrument itself.

Do not fall into that category tonight. Who's first to see the Medici stars?'

One of the merchants stepped forwards.

'Don't touch the tube in case you move it. The alignment must be precise,' said Galileo.

The other guests waited in silence as the man squinted with first one eye and then the other. In between attempts, he cast a wary glance at the crowd. Eventually he straightened up, blew out a deep breath. 'Sorry, signor. I think you forgot to put the crystals in this one.'

Galileo checked the alignment. 'But they are clearly there. Giovanni, your turn. Please, restore some sanity to this gathering.'

Magini took a long pull on his wine. 'I am sceptical about even putting my eye to the device.'

'Don't tell me you are going to refuse to look. I thought that foolishness was the preserve of old Libri.'

'Show some respect. Libri is gravely ill, perhaps even on his deathbed.'

'Then let us hope that having failed to see the Medici stars in life, he gets a good view of them en route to Heaven.'

Magini tutted. 'And why call them the Medici stars?'

'I am Grand Duke Cosimo's court astronomer and philosopher. What could be more natural than to honour my patron?'

Magini scowled.

'I will look through, and put an end to this bickering,' Cesi announced. He took up his position and stared for a long time. Anxious looks passed between the onlookers, then a few whispered comments. Cesi remained fixed at the tripod as the level of conversation grew around him. He gave a little nod and straightened his back. The men hushed. Galileo stepped forward. 'Well?'

'I see them,' said Cesi. ' Exactly as Galileo describes.'

There was a cheer that lifted Galileo almost to the stars himself. After that the guests trooped up to look through the tube. Some laughed. Some sighed. All of them left the tube with shakes of their head and looks of astonishment. Galileo soaked up their compliments with what he hoped was due modesty, a swell of vindication growing inside him. Yet he did not see Magini take a look.

No matter. Who is the greatest astronomer now?

At one point Galileo noticed Grienberger behind the tube. Intrigued, he excused himself from a conversation and went over, arriving just as Grienberger straightened up. 'How do my optics compare to yours?'

'They are broadly comparable.' The wide face was not giving anything away.

Cesi called them all to order. 'Tonight, we are gathered to honour our great guest,' he held his hand towards Galileo and a round of applause broke out. 'Tonight he becomes one of us, a Lyncean. Having read his *Siderius Nuncius*, I wish to extend this pledge. From now on, Galileo, you will never have to search for a publisher. The Lyncean Academy would be honoured if you were to allow us to publish all your future works, so that everyone can benefit from your wisdom. We will pay for the production and ensure that the books are widely read.'

Galileo was touched by the commitment. 'Thank you, Prince Cesi, I accept.'

'Then I need say nothing more except to call upon Professor Demisiani to make a very special . . . suggestion.'

A rotund man stepped forward, his face ruddy in the jittery lantern light. When he spoke, his large teeth shone white. 'If I may be so bold, Signor Galileo, I wish to propose a name for your incredible device. In my way of honouring the ancient Greek astronomers, I propose to take *tele*, meaning "afar", and *skopeo*, "to look at" and bring them together to give us *telescope*, meaning "far-seeing". I humbly beseech you to accept the name as our gift to you.'

Cesi bit his lower lip and dipped his head in Galileo's direction.

Galileo was confused by the logic behind honouring the ancient Greeks for his discoveries, but the expectation on Cesi's face convinced him of the right thing to say. 'Sir, while I point out that the naming of the instrument is not mine to adjudicate, I am honoured to be consulted. And let me say that I humbly accept your gracious offer of membership to the Lyncean Academy. We share a common set of goals and beliefs. Namely that philosophy is written in this grand book of the universe and stands continually open to our gaze. But the book cannot be understood unless one first learns the language in which it is composed. It is written in the language of mathematics; its alphabet is triangles, circles and other geometrical figures. Without these it is humanly impossible to understand a single word of it, and one is left to wander about lost in the dark labyrinth of the sky.'

With the formalities completed, the partying recommenced. Galileo slipped away from the crowd and found himself a large laurel bush to stand behind while he emptied his full bladder. As he buttoned himself up afterwards, he heard voices. Cesi and Grienberger. He froze.

'I am here to caution you on behalf of the Roman College,' the Jesuit was saying. 'You must be careful in your advocacy of Galileo.'

'Why? He is the greatest astronomer alive!'

'We cannot jump to conclusions about the meaning of these new planets. It goes too much against Scripture.'

'But you have seen them with your own eyes.'

'The Roman College is the arbiter of Catholic science, and we are favourably disposed towards Galileo's discoveries, but everything must proceed along established lines. We will report to the cardinals, who will report to his supreme eminence, the Holy Father, but we need support among the theologians if we are to argue that the Bible needs reinterpreting. We are doing everything we can to get the consensus we need. In the meantime, proceed with caution: stick to the facts of the discovery but mention nothing of interpretation. If you follow my advice,

Prince Cesi, the Jesuits will remain your friends.' There was an edge to Grienberger's tone that chilled Galileo.

'Of course.' Cesi's voice sounded like that of a scolded child.

'Good. Let us return to your guests.'

Shaken by what he had heard, Galileo waited until he was sure they had gone before rejoining the party. He took another glass of wine but his mood was broken, and he soon asked if he might leave, feigning tiredness. Cesi escorted him back down the hill, where Grienberger was waiting by the carriage. Galileo's heart missed a beat at the sight of the Jesuit.

'Galileo,' said Grienberger in his matter-of-fact tone, 'you are to be ready at ten o'clock, tomorrow morning. I will meet you at the Tuscan embassy with a carriage. Do not be late.'

'Who it is that summons me?'

A quizzical look crossed Grienberger's face. 'Do you really need to ask? You have been granted an audience with His Holiness.'

If Galileo thought he had been nervous about visiting the Roman College, he now realised his anxiety had been nothing compared to the terror of standing at the gates of the Vatican. His insides bubbled as his eyes fell on the great central dome and its flanking towers rising up from the rooftop of St Peter's Basilica. Last night's eavesdropping had unsettled him. His discoveries had never been intended to undermine the Scriptures. He had not anticipated that they could be seen as an attack. His lack of foresight added to his nerves. *What else have I failed to anticipate?*

The Vatican was where the cold majesty he had seen in the city gave way to extravagant craftsmanship. And it was becoming increasingly extravagant. The white stone of the building's frontage was covered in wooden scaffolding and swarming with workers. Through the web of poles and planks, Galileo glimpsed a set of grand columns rising to support a broad triangular pediment.

Beside him was his shadow: Grienberger. Even under the direct glare of the sun, the man's face betrayed no emotion. There was no hint of discomfort at the heat, just that hangdog expression and the annoying tendency to never quite meet Galileo's eyes. *Should I be fearful of him or grateful for his presence?*

'It is unusual of His Holiness to take such an interest in the mathematical sciences. You are truly honoured.'

'I know,' said Galileo, aware that his entire credibility could be destroyed if he uttered a single wrong word. The Pope's word was law. He could overrule Galileo in a heartbeat and no one would listen to him again. But if he were to endorse Galileo . . .

'I have one instruction,' said Grienberger. 'Do not offer to show him the telescope. If His Holiness wishes to see the heavens, the Roman College will arrange the viewing.'

Galileo meekly agreed.

Inside, the high ceiling rested on graceful arches that reached down, like a giant's shoulder supporting a mighty burden. Every square inch of wall and ceiling space was decorated with frescos, geometrical in pattern and depicting biblical portraits and scenes. Galileo was so lost in his admiration for the building he almost missed the approaching scarlet-clad figure.

'Cardinal Bellarmine, this is Galileo Galilei,' said Grienberger.

'May I welcome you to the Vatican.' The cardinal's expression was neutral as he greeted Galileo.

'I haven't seen you at the College lately,' said Grienberger.

'If only I had the time to visit; I swear the Inquisition becomes busier every day,' said Bellarmine.

Galileo's heart beat faster. *The Inquisition? Why were they involved?*

'But we are winning the fight, yes?'

'Indeed we are. The Lutherans are becoming increasingly isolated.'

'What hope of England?'

'What indeed? It is a dark place. The failed assassination of James has done us no favours. It has hardened his resolve to outlaw our loyal Catholics, however innocent they may be. We are far from resolution there. But let us not wallow in such matters. Today is a celebration. Allow me to show you to your audience.' Bellarmine betrayed his years only when walking, emitting occasional grunts of pain. 'Arthritis,' he explained. 'I used to be able to count on it easing in the summer.'

'My sympathies, Cardinal,' said Galileo, hoping to extract some warmth that he could take as reassurance. 'I suffer too, though this year has been merciful to me.'

Bellarmine led them from the magnificent arches into a more bureaucratic corridor, though one still adorned with busts between the windows and a floor of black-and-white diamond-shaped flagstones.

Two of the Pope's Swiss Guards stood at the end of the corridor. Uniformed in stripes of orange, red and blue, they opened a pair of wooden doors in unison, allowing Galileo's party to pass without breaking step. Pope Paul V sat at the far end of a wide strip of scarlet carpet.

His old face might as well have been carved on one of the busts that Galileo had walked past, moving only with a small nod when an aide in cardinal's robes bent to his ear and whispered something.

Bellarmine and Grienberger dropped behind Galileo, who walked forwards at what he hoped was a dignified pace and knelt before the Pope, touching his lips to the papal ring on the hand that the Pontiff held out with casual indifference.

'I am told that you have seen wonders hitherto unanticipated in the heavens.' His voice was warmer than Galileo expected.

'I have been blessed in this way, yes, Your Eminence.'

'What is it you see?'

Galileo described the moons of Jupiter and the myriad stars he had seen in the Milky Way. He was about to launch into his discussion of the way the mountains on the Moon proved it was another Earth, when something stopped him. He decided to simply mention them as markings.

'Heaven is richly stocked with new wonders. I cannot help but speculate as to what else I may find.' Immediately the words were out of his mouth, he wondered whether he had gone too far with that presumption.

'I have one concern,' said the Pope.

Galileo's stomach clenched.

'Why would God hide these marvels from us in the first place?'

'Perhaps to urge us to greater heights of achievement than any of us thought possible.'

The Pope pursed his lips for a moment and then rocked his head in agreement. Galileo risked looking out of the corner of his eye. There was a change in Grienberger's face. He was not quite smiling but the corners of his mouth had definitely lifted.

The Pope spoke again. 'Galileo of Florence, you are a credit to the Catholic Church. God go with you in your work.'

Nothing had changed in the buildings, the sounds or the high summer smells of the city when Galileo emerged from the Vatican, yet as he looked around, everything felt different. He wanted to laugh or maybe cry; it was difficult to tell which.

'You did well today, Galileo. You have learned much in your week with us here,' said Grienberger.

Ordinarily Galileo would have taken offence at being patronised by a younger man but today he found it easy to accept the words as a compliment. 'Thank you. I have a question for you.'

Grienberger inclined his head.

'Do you not think that the weight of my observations proves Copernicus?' asked Galileo. 'The fact that the Medici stars orbit Jupiter, not the Earth; that the Moon has earthly features . . .'

'The official Jesuit position is currently an Aristotelian one.'

'But my observations are so clearly in conflict with Aristotle.'

'Listen to my words, Galileo: "currently". As Father Clavius said to you, the Jesuits recognise that the orbs of Heaven need rearranging, but as yet we have no new interpretation to offer that satisfies everyone.'

'I didn't ask you what the Jesuits think, I asked what you think.'

'Galileo, I do not enjoy the same freedom as you. The Jesuits speak with one voice. Each of us understands that the individual is subservient to the group. There are 13,000 of us, some as far as China, all spreading the word of Christ with the same voice.' There was obvious pride in Grienberger's voice, the first time Galileo had perceived any passion in the man. 'We achieve this unity because we work with a common purpose, with clear lines of communication and an acceptance of the hierarchy. Everything we publish passes through the College and is reviewed before it goes near the presses; in this fashion we maintain the quality of what we do. It is our organisation, our unity of purpose that makes the Lutherans fear us.'

Galileo turned to face him. 'These new observations can help in the fight, but only if we are willing to use them. We can show people that the true glory of Heaven is a Catholic revelation.'

Grienberger's head shifted a fraction. For a moment, Galileo actually saw into the Jesuit's blue eyes. 'I agree, Galileo, but you need to work with us.'

'One voice,' said Galileo after a pause.

'One voice. It is what makes us strong. Know this, Galileo: stick to describing the basic facts of your observations, and let us do the rest. Do this, and you will have nothing to fear from us.'

'And if not?'

Grienberger ignored the question.

21
Prague, Bohemia

The entrance hall was full of linen. Bundles of sheets lay on the floor; others were scattered higgledy-piggledy across the furnishings. As Kepler looked at the mess, some more came flying through the kitchen door, followed by pulses of angry conversation.

He headed for the source of the trouble, his face grim. Barbara stood with her hands on her hips, shrieking at the top of her voice. Frau Bezold had her back turned, pretending to ignore the tirade and folding one of the bedsheets.

Barbara's face was bright red. 'Husband, tell her! We cannot sleep on dirty sheets.'

'There's nothing wrong with my laundering,' said Frau Bezold.

'You must do them again until they are cle—'

Barbara dropped to the floor.

She fell so quickly that it took a few seconds for Kepler to comprehend what had happened.

'Don't just stand there!' urged Frau Bezold. But no sooner had the housekeeper taken a step towards her mistress than she recoiled, her face twisted in horror.

Barbara was thrashing wildly on the floor. Her limbs made the most sickening thudding noise on the flagstones. Livid bruising was already visible on her elbows and wrists.

Kepler clambered around the table and dropped to his knees, trying to grab hold of his wife's flailing arms. But there was too much force in her jerky motions. Every time he thought he had her, she would break free with a renewed surge of wild energy.

'Barbara, Barbara,' Kepler called. 'It's me, Johannes.'

Her head may have moved at his voice, but it was difficult to tell.

Frau Bezold began to recite the Lord's Prayer.

'Help me, Frau Bezold, we must restrain her.'

She clutched her crucifix and did not budge.

'Help me, now! There is no danger, except of her injuring herself.'

Frau Bezold gingerly approached and, with a wail, reached out to grasp Barbara's ankles. Together, Kepler and Bezold fought to keep Barbara still. Her eyes were half-open but unfocused. Then, as quickly as it had begun, so it ended. Barbara went limp and seemed to fall asleep, emitting a guttural snoring. Kepler hesitantly released the slumbering form. Neither he nor Frau Bezold spoke but he began to tremble, so leaned forwards on his knees.

When Barbara woke up, she was confused and weak. She reached up and clutched her husband, who manoeuvred her into his arms. She dug her fingers into him, twisting great handfuls of his clothing.

'You're alright now. You're safe. Let's get you into bed,' he said.

There were pale marks on Barbara's neck where the doctor touched her carotid artery. Kepler stood beside Frau Bezold and watched the man from the corner of the room.

The doctor looked over his shoulder at him. 'Seizures, you say?'

'A handful in the past few days.'

'A handful? What do you mean? Three? Four? Five?'

Kepler felt like a child, caught concealing the truth. 'Five in three days, Doctor Reichard. But she has not been herself for many months now.' He flicked a glance at Barbara. 'She suffers from melancholia.'

Reichard closed his eyes and counted Barbara's pulse. 'A little high, but nothing to worry about.'

Kepler smiled at Barbara, who responded weakly. She was sitting upright, immobile, with the blankets pulled high around her. Reichard knelt down and ran a hand under the bed. 'Aha.' He retrieved a ceramic chamberpot and looked into it, fascination written on his face.

He took the pan to the window, tilted it towards him, then away. There was a faint slopping sound. He swirled the pan and held it to his nose to sniff, as Kepler had seen Hewart von Hohenburg do with wine. He reached into his bag and retrieved a small glass beaker. Into it, he splashed a sample from the chamberpot and held it to the light at the window. With a noise of satisfaction, he tipped the contents back into the pot and dried the interior of the beaker with a cloth.

'Doctor?' Kepler asked.

'Nothing much wrong there. But, melancholia, you say? That is linked to an excess of the cold and dry vapours in the blood. The best course of action will be to bleed her at the temporal vein.'

Kepler nodded. Barbara remained motionless.

Reichard retrieved a small knife from his bag. He ran a fingernail down the dull blade, scraping at a collection of black spots on the metal. Then he pulled out a metal bowl.

Kepler noticed how reluctantly he approached Barbara and reached towards her right temple. She winced at the incision and screwed her eyes shut as the blood flowed from the side of her face. At arm's length, the doctor held the bowl until sufficient blood had pooled in the metal receptacle, after which he pressed a cloth to the wound. Frau Bezold immediately rushed to hold it in place as Barbara slumped backwards onto the pillows.

'She will sleep now.' He turned to Kepler. 'Can we talk, privately?'

Once outside, Reichard took a deep breath. 'Herr Kepler, I'll come straight to the point. You do not need a doctor, you need a priest. The bleeding will help the melancholia but the seizures . . . It is beyond physic . . .'

Kepler squared his chest. 'I cannot believe it.'

'Herr Kepler, please. You must realise – evil is afoot in Prague.' The doctor took a deep breath. 'I can find nothing medically wrong with your wife. The only explanation is that she has been . . . invaded by an evil spirit.'

'You are wrong.'

'You saw her thrashing, was it not demonic? She is fighting a war inside. You owe it to her – and to the rest of us – to stop this terror spreading. I urge you to help her with an exorcism.'

When the doctor left, Kepler brooded on the diagnosis, finding reasons why it could not be true. Yet that night he arranged the children around Barbara's sleeping form and conducted them in an hour of prayers and recitals. He pronounced each word more carefully than he could ever remember doing in the past, emphasising their meaning by holding each child's gaze in turn.

In the morning, Barbara awoke with a smile, blinking at the brightness from the window.

Kepler sprang up from the armchair in which he had spent the night dozing. 'You look better.' He had barely completed the sentence when her eyes rolled upwards and another violent fit took possession of her.

Kepler made his way through the deserted streets of the Jewish quarter. Arriving at the university, he found Jessenius bustling down a corridor.

The anatomist had his hands clasped behind his back and appeared to be studying his feet. When Kepler intercepted him, his head jerked upwards.

'What are you doing here, Johannes?' It sounded like an accusation.

Kepler explained about Barbara.

'I am sorry to hear that.' Jessenius looked drawn.

'Tell me you do not believe in this, too.'

'I would prefer such things not to be true but they make sense to me.'

'How? How can it make any sense?'

'There is one thing that strikes me whenever I perform an autopsy. It is that by the time the body ends up on my slab, whatever animates the flesh has long since fled.'

'You mean our soul?'

'Yes, implanted in us by God forty days after conception and set free at the moment of our death. The soul controls everything: our growth, our perception, the way we move and the way we think. Without it we are nothing. And just as tenants occupy houses, evil spirits can invade our bodies.' Jessenius scanned the corridor.

'Am I keeping you?'

'No, no, no. Let us keep walking. Where was I? Oh yes, think of it this way. The liver creates the blood, which flows to the brain where the nervous spirits are created. These are then transported by the blood around the body and control our movements. We accept that illness is caused when vapours contaminate the blood. These slow us down, make it difficult to carry out our daily lives. Seizures, on the other hand, are completely different. The movements they cause are frightening, are they not?'

The sight of Barbara writhing on the floor flashed into Kepler's mind.

'Seizures look to me as if the body is fighting against something malevolent that wants to take control of it.'

'I cannot believe that evil dwells in my wife.'

'Johannes, think: has she been bad-tempered lately? Not herself? Melancholic? All these things make her susceptible to bad spirits. You look sceptical. Ask yourself this: are you reluctant to believe your wife is fighting evil because you don't believe in spirits, or because you don't want to believe that someone you love has become their victim?'

Hasty footsteps drew their attention. An agitated man approached. Jessenius glanced sideways at Kepler then at the new arrival. 'Well?'

'They're not after us. Their argument is with Rudolph. They say they've not been paid, so they'll take what they want from the city.'

Fear etched Jessenius's face. 'We can't just stand back and let them invade Prague.'

Hewart von Hohenburg's warning about Rudolph's out-of-control mercenaries rattled inside Kepler.

'Are the men ready?' asked Jessenius.

The messenger nodded.

'Very well. Send the riders out. They are to find Matthias and ask for his support. We'll hold the city for as long as possible.'

The messenger hurried away, leaving Jessenius breathing deeply.

'Jan, what are you involved in?' Kepler said.

Jessenius looked up, ashen. 'I am not just an anatomist, Johannes. I've used my position at court to negotiate on behalf of the Protestant estates with Rudolph, but he's betrayed us with these Bavarian mercenaries. We have been talking to Matthias and will now side with him. With Matthias as Emperor, we can stabilise the city again.'

'But the bloodshed . . .'

'Go home, Johannes. Lock your doors. The golden age of our beautiful city is about to end.'

The streets were empty. Kepler could see twisting columns of black smoke rising in the distance, but there was no noise yet of the advance. Not a soul was in the market square; a few abandoned stalls and some squashed fruit and vegetables were all that remained. Kepler felt horribly exposed. He darted into the shadows and skirted along the walls.

A hand fastened onto his arm, pulling him so sharply that he lost his balance as he was dragged into an alleyway. He lashed out against the vice-like grip and tried to get his feet back under him, but it was to no avail. A hand that smelled of earth clamped over his mouth.

'Not that way, friend. Not if you want to live,' said a gravelly voice next to his ear.

Kepler stopped struggling. His captor was a man of the countryside: sunburned cheeks and forehead, calloused hands and creased eyes. He let go of Kepler, who peered around the dark alleyway. It was crowded with men, the smell of their bodies pricking his nose. Kepler saw faces of all ages in the shadows. Irregulars. Workers from the surrounding estates rallying to defend their city.

The man who had grabbed Kepler drew a dagger from his belt and flipped it so that the handle faced the astronomer. 'Take this, you're going to need it.'

Kepler backed away from the weapon.

'Suit yourself, but Rudolph's men will be all over the square and they won't care if you're armed or not.' He replaced it in his belt.

'I have to get to Karlova Street,' said Kepler.

'You'll run straight into them. Get down!' He yanked Kepler back into the shadows. The sound of trotting hooves began to echo around the stone buildings. Kepler's breathing quickened as he watched twenty, thirty, fifty, a hundred horsemen pouring into the square. He lost count; they just kept coming.

Behind him, Kepler could hear muttered prayers.

'We wait for the signal,' hissed someone.

A horse trotted so close to the end of the alleyway that Kepler could see the dark stain of blood smeared across its rider's sword. The man rolled his shoulders, flexed his head from side to side, and spurred his horse onwards. Another horseman filled his place, then another and another.

The crack of a single musket shot ricocheted around the market square. Kepler was engulfed in noise. The men hollered at the top of their voices and charged out of the alleyway, driving Kepler onwards with them.

The chances were that the cavalryman never knew what hit him. The same hands that had manhandled Kepler dragged the soldier from his horse. As he toppled, a blade slashed a ribbon of blood across his neck, and his twitching body fell to the ground. The swiftness of his death terrified Kepler; somehow it seemed both trivial and momentous.

More irregulars joined the battle, swords waving, from the many side streets. The cavalry were trapped, pressed so close that there was no room to manoeuvre their horses. Each swing of a horseman's sword was as dangerous to their fellows as for the men they were fighting. In the midst of the crush, a horse reared, hurling its rider backwards into an abyss of flailing hooves.

A man took aim and fired a musket into the crowd. There was a bloody explosion as the ball found its mark. Then the marksman himself crumpled, run through by a horseman. The rider hollered in triumph and signalled for his men to follow him in a charge towards the main skirmish.

Wounded men screamed, but the worst cries were from the horses. Slashed or shot, they fell, adding their own agonies to the bedlam in the market square.

Kepler tried to block out the excruciating screams and the sharp report of musket shots, the tinny clash of swords and the dull thud of lead balls into soft flesh. The battle had become an indefatigable engine of hacking, bludgeoning death. Man after bleeding man toppled. The metallic tang of blood was overwhelming, creating a fetid invisible mist. A panic-stricken horse careered into Kepler, and the collision sent him sprawling onto the blood-soaked ground. He found himself among the bodies. A man, face streaked with dirt, lay on the ground nearby, silently weeping. Kepler rolled him over to cradle his head, and watched the light fade from his eyes.

Then Kepler saw it.

Scoto's booth, toppled but intact. He clambered over the cadavers and crawled inside the wooden shelter. Inside, he curled into a foetal position, closed his eyes and pressed his hands to his ears.

22

He had no idea how much time had elapsed. All he knew as he emerged from the shelter was that the worst sounds of the battle had gone and his quaking had subsided. There were still noises coming from the north, in what could be the Jewish quarter, but that was far enough away not to be an immediate threat.

The dead lay everywhere. The hot smell of blood and guts and emptied bowels fouled the summer air. Already, a few old women and children were picking over the corpses for rings and other valuables. One old woman looked up defiantly as Kepler drew near. Heaving erupted inside him. He bent over to vomit. He retched time and time again until his stomach was completely empty.

Tearfully he straightened up and took a few faltering steps. The old woman regarded him with disdain; she reminded him so much of his mother that he almost addressed her as such. He opened his mouth before realising the absurdity of it. Katharina would be safe, at home in Leonberg. Ignoring the hag's sneer, he stumbled away.

That's when his crying truly started: huge sobs of grief for those slain today, and those slain yesterday, and for those whose time would be forced upon them tomorrow. All of it sparked by Rudolph's ineptitude in government. The blubbery fool spent his days hiding in the Kunstkammer, weeping over the beauty of some painting one moment and then issuing murderous orders to look good to the Pope the next.

Kepler had seen the Jesuits at court, gliding through the corridors with their stern expressions masking whatever dark mischief they were there to orchestrate. They would march in on Rudolph, who seemed powerless before them, swaying like a sapling in the breeze in his desperate attempt to cling on to power.

All of the Emperor's previous treaties became as easy to break as one of Susanna's daisy chains, and, as a result, a Bavarian army was rampaging through his city, slaughtering his people. Despite Rudolph's education, wealth and upbringing, Kepler had never known a more ignorant man. He doubted Rudolph knew what was happening outside. Even if he did, it would mean nothing to him. He would claim it was the work of God, cleansing the land of heretics. God! This was nothing to do with God. This was all too human power play for earthly authority. God must surely find this bloody spectacle repellent.

When would people be free to live their lives, worship God in their chosen way, and no longer fear the onslaught of some foreign army? Kepler angrily rubbed his eyes to clear his vision. *Maybe never*, he thought, *maybe never*.

There were a dozen or so bodies in Karlova Street, ordinary citizens trampled by the horses. At first glance, none of them looked familiar. Then Kepler saw, face down in the gutter, the broken body of the man who used to sharpen knives.

Kepler reached home. Thankfully there were no broken windows or signs of fighting. He found the front door locked firmly. The back door was also apparently secure, but an experimental push produced movement and the scrape of a heavy piece of furniture. He shoved harder and forced the door open far enough to squeeze through. Frau Bezold must have shoved the kitchen table against it.

There was a pail of water on the floor. Kepler plunged his hands into it and splashed his face, needing the sting of the cold water to prove that he was still alive. The more he repeated the action, the dirtier the water became, stained by the dirt and gore that had clung to his hands.

A whisper of movement made Kepler turn. Frau Bezold was inching her way into the kitchen, shoulders hunched, holding the family carving knife. She was squinting so tightly, her eyes were practically closed.

'It's me,' said Kepler.

She relaxed, dropping the knife. 'It's the Master,' she shouted.

'Come and sit down,' said Kepler.

She was shaking as he guided her gently towards a chair.

Susanna came scurrying down the stairs to meet him near the kitchen door. 'Papa, hurry. It's Friedrich.' She grabbed his hand and pulled him urgently up the stairs.

Barbara was sitting on Friedrich's bed, holding him in her arms. She was clutching him so fiercely the boy looked in danger of suffocating.

'What's wrong with him, Barbara? Barbara?'

She rocked back and forth, oblivious to Kepler's presence. A terrible thought occurred to Kepler. One of dark forces manipulating Barbara to spread their malignancy.

'Let go of him, Barbara.'

She favoured him with an evil stare, increasing his resolve.

'Let go of him. We must let him breathe.'

Kepler prised Friedrich from her arms.

'It's like the last time . . . when we lost Heinrich,' said Barbara.

'No, Barbara. Heinrich was a baby. Friedrich is strong and healthy.'

Barbara returned to her rocking motion at the bedside as Kepler stripped the boy of his outer clothing. He checked Friedrich's hands and feet for the marks of the Hungarian plague. Thank God, there were no such blemishes but there was an angry rash across the boy's torso. He was anxious to move Barbara away.

'We can leave Friedrich to sleep now. Frau Bezold will look after him.'

Barbara shook her head. 'I'm not leaving him.'

Kepler began to protest when Susanna distracted him. She was standing next to the doorframe, her face flushed. 'Papa, I don't feel very well, either.' He rushed to her and conducted a similar examination. She too, had a rash.

By midnight, Friedrich's rash had transformed into a multitude of spots.

'Don't scratch, you'll make it worse.' Kepler eased the boy's arm away from his face.

Frau Bezold found a tincture of camomile and some clean cloths. Together, she and Kepler dabbed at the children. In the third bed, Ludwig too was showing signs of the illness.

Barbara had not been taken by a seizure all afternoon or evening. That single fact alone was all that comforted Kepler. She slept now in the chair by the side of the children, her head lolling.

During the night, Kepler prowled the house, watching from the windows. Every shout from outside seemed amplified, every distant fusillade a threat. He traced the glow of outlying fires, orange as the sunset, above the surrounding rooflines. Incandescent cinders wafted into the air and Kepler fancied that each was a human soul beginning its flight to Heaven.

Frau Bezold catnapped in a chair outside the children's room.

'How are they?' he asked during one pass.

'Settled now. What's wrong with them?'

Kepler looked down at his own scarred hands.

'Smallpox.'

By dawn, the city felt calm. Head pounding and limbs heavy, Kepler stepped into the street. Black birds circled in the sky. Dogs loped by, noses down. Most of the bodies still lay where they had fallen.

Unwilling to see the market square again, he cut down an alleyway and emerged in a neighbouring street where the crows were busying themselves on the human carrion. He chased the birds into the sky, hollering at the top of his voice.

Somewhere across the city a bell began an empty toll.

Kepler arrived at the Jewish quarter. Many of the houses lay in confusions of blackened beams and blistered timber; some still smouldered. Everywhere were twisted bodies, as charred as the homes they had once occupied.

A small girl wandered into view, obviously searching for someone. Kepler knew better than to approach her. Already men were gathering in the few remaining doorways to cross their arms and glare at him. He risked a tentative step towards one of the householders. 'Who did this to you?'

'Your kind.'

'You mean the mercenaries?'

'No, the irregulars. The mercenaries ran. The Protestants couldn't keep up so they did this.'

Kepler stared again at the devastation and tried to match it with the words he was hearing. The Protestants were defending the city; they were not here to sack the Jews. Yet as he was shaking his head in disbelief, he remembered the confusion in the market square, the inhuman things he had seen there. He could hear it again, smell it again – the horror of the bloodlust. Tremors of betrayal rippled through him. *How could they have done this to innocent people?*

'You must be mistaken,' he pleaded.

The man took a purposeful step towards him. 'No mistake. And many here want revenge.'

Kepler turned and fled, screaming his rage into the streets. *What was one more cry of anguish in a city that had echoed to so many just a day before?* He passed the university, his original destination, and headed straight for the bridge, his sights now set on Rudolph's palace.

Nervous soldiers guarded the entrance. A captain recognised Kepler and waved him onwards. As the gates opened, Kepler saw Tengnagel astride a skittish horse, waiting to leave. Tycho's son-in-law noticed Kepler too but did not acknowledge him. He spurred his steed and galloped out into the sunlight, head high, followed by a small mounted entourage who struggled to keep up.

Inside the Palace was a cacophony of arguing officials. Nothing could be decided in this commotion, and Kepler wanted to scream at all of them. When he finally found von Wackenfels, he saw that the Privy Councillor was on the verge of tears. His blond hair stood on end in greasy stubs. The skin under his eyes was shadowed and there were ink stains on his fingers.

'Johannes, thank the Lord you are here. The Emperor is asking for you. I was about to send a carriage. Come this way, we can reach him faster if we use the lower corridors.'

Von Wackenfels led him away from the hubbub and down a narrow staircase, almost concealed behind a stone pillar. There was no daylight in the lower passage. It relied on illumination spilling out from the various chambers that hung like ribs from its backbone.

There was a chemical tang in the air. As they hurried by an open chamber, Kepler could not help but stare.

Inside, smoking trays were suspended over candles. Towards the back of the room an alchemical furnace roared. In the hellish heat a man was on his bended knees, praying in front of a tabernacle. His long white beard was waxed to a point and he wore clothes that resembled a priest's. However, instead of hands clasped in Christian prayer, he held his arms in a wide beseeching gesture as he dipped his forehead to the floor.

'Magic?' asked Kepler despairingly.

'The application of philosophy. What good is knowledge if we cannot tap into it and use it?'

'You cannot believe that we can summon spirits?'

'Why not? The priests summon spirits.'

'No, they don't . . .'

'I believe what His Majesty believes.' There was a flash of temper in von Wackenfels's voice. His face quickly softened but remained drawn. 'That's my job.'

They found Rudolph in one of the nearby chambers, huddled on the floor. Drawn in chalk around him was a circular band, divided into twelve. Within each section was a scrawled symbol of the zodiac. Beside Rudolph was a tapered spike, some three feet long and fashioned out of some flaking mineral. In his hands, Rudolph caressed a golden cup.

Von Wackenfels put his arm across Kepler's chest. 'Do not approach too closely. You must not enter the magic circle.'

'What's he got in there?' Kepler whispered.

'His two most precious possessions: a unicorn's horn and the cup of Christ.'

Kepler's anger threatened to burst from him. He wanted to storm into the circle and grab the Emperor, drag him down into the streets and throw him onto the pile of bodies. That way, perhaps he would understand.

'Your Majesty, Johannes Kepler is here,' said von Wackenfels.

Rudolph began his usual mumblings. As the litany continued, all Kepler could make out was the word 'horoscope'.

'I will not give you a reading,' he said.

More incoherence, out of which curled, '. . . do as I ask. I must know how this silly skirmish will end.'

'*Silly skirmish?* Your citizens are being slaughtered. The blood lust runs through both sides.'

'What can I do? What can I do? I must know. You must tell me.'

Von Wackenfels stepped forward. 'The estates have called for Matthias's help. His troops are marching towards the city as we speak. They are well disciplined and a force to be reckoned with. Tengnagel has gone as an envoy for His Majesty. Please, Johannes, how will this end?'

'I don't need to draw star charts to tell you this. Pay the mercenaries to leave the city. Once they have gone . . .'

'Pay them! Pay them!' Rudolph tottered on the edge of the chalk marks. 'For looting my city? Pay them, you say?'

Kepler's heart beat faster. 'Your Majesty, they're here to claim their money. The only way to be rid of them is to honour your agreement to pay them. Once they have gone, you will have removed the need for your brother's intervention, and he will stop his advance.'

Rudolph started giggling. Soon, his mirth consumed him so utterly that he lost his footing and dropped to the floor. Even this did not halt his amusement. Von Wackenfels squirmed at the display, and Kepler realised that whatever rationality had once resided in Emperor Rudolph was now gone for ever.

23

Florence, Tuscany

Sunlight bathed Galileo's rooftop terrace. He basked in the warmth and the faint smell of citrus from the lemon trees as he pumped his foot up and down on the lathe's pedal. The spinning wheel kept a steady rhythm and the clicking cogs added their own syncopation. He dipped the stumpy polishing tool into the finest of his grinding powders and held it to the glinting lens fixed to the spinning wooden armature.

From up here, Galileo could keep one eye on his task and the other on the narrow road that ran alongside his property. Although he was cloistered in a quiet part of the city, the occasional activity near his villa helped to combat the laborious nature of the lens polishing.

If he lifted his eyes further he could see the tall cypress trees that punctuated the terracotta wash of the city's buildings.

'Signor?'

Galileo took his foot off the pedal and dropped his gaze to the street. A rotund monk with an equally rotund face was looking up, his hands raised to shadow his eyes. Galileo smiled. 'Well, well. Brother Benedetto Castelli. Don't tell me they've thrown you out of my old job already?'

'This is no time for jokes, Master,' panted Castelli. 'I have an urgent matter to discuss with you.'

'Not run out of food in Padua, have they?'

'It's about Copernicus.'

The smile faded from Galileo's face. 'I'll come and let you in.' He pulled off his leather apron and went down the twisting staircase to the garden gate.

Castelli was much younger than he appeared from a distance. He had dark eyes that shone as brightly as one of Galileo's lenses. His eyebrows and moustache were as brown as a forest bear although his hair was receding, rendering a tonsure superfluous. His large fleshy ears glowed red and beads of sweat clung to his forehead.

'Come in, before you fry in all your fat. Lemonade?'

'Anything to quench my thirst. How are the girls?'

'Settled in at San Matteo now, just down the road. Still too young to take the veil, of course, but that gives me some time to find them suitors.'

Castelli wiped his brow. 'How goes the search?'

'I'm run off my feet making telescopes. I haven't even given it a thought. It's cruel that Virginia was born a woman, she's so quick and curious about my work she would have made a formidable philosopher.' Galileo ushered Castelli inside. The sitting room was adorned with half-made telescopes and lenses. They navigated the open pots of glue and other obstacles, and headed for the kitchen.

'You didn't tell me that the faculty at Padua were so fanatical in their dislike of Copernicus.'

'No more than anywhere else,' said Galileo.

'I was warned by the overseer never to teach or to mention the movement of the Earth. The subject is forbidden.'

'Did you run all the way up the hill just to tell me this?' Galileo began clearing a space on a chair.

'No,' said Castelli.

There was a clatter. Galileo turned to see that the tubby monk had bumbled into a low table, tipping candles and quills across the floor. Stepping back, Castelli then sent flying a pile of papers covered in calculations.

'If I'd known you were coming, I'd have cleared out the furniture.' At least having Castelli around made Galileo feel graceful again.

'You should make your housekeeper tidy up,' said Castelli.

'I don't have one. She kept tidying up. Couldn't find a thing.'

'So who makes the lemonade?'

'Not that it is any of your concern, but Virginia. She is proving a godsend for my laundry too. She takes it from me when I visit and has it all washed and pressed a week later. Now, what have you come to talk to me about?'

Castelli picked up a messy sheaf of embossed leather and carefully lowered himself onto the chair beneath. 'Opposition is growing to the Copernican way of thinking.'

'Is that it? Just give it time. The more telescopes I build, the more people will see the truth.'

'No, you don't understand. It's not about disbelieving what you see through the telescope any more; it's about disbelieving that the Earth moves. Some are saying that it is reckless talk and goes against the Bible.'

'What fools are they? Tell me names and I shall make a mockery of them.'

Castelli screwed his face up at Galileo's words. 'No, sir. It's not a man; it's a woman.'

'Ha! Even easier.'

'Galileo! It's Madama Cristina.'

'The Grand Duchess? I don't believe it. Who have you been talking to?'

'The Grand Duchess herself.'

Galileo froze. 'Tell me exactly what happened.'

'I was dining with the Medicis in Pisa two days ago. I was regaling them with talk of the telescope and the movement of the Medici stars. The Grand Duke showed himself to be much pleased with everything I had to say, but Madama Cristina asked how I could be certain that the stars were real and not illusions of the telescope. I described to her the orderly movement, and the Grand Duke agreed with me that there could be no doubting their reality. But when I was leaving, a porter came running after me and called me back. Before I go on, I should tell you that Cosimo Boscaglia was there.' Castelli looked almost apologetic.

Galileo groaned at the mention of the aged philosopher.

'He was sitting next to the Grand Duchess and had her ear for long periods during the meal. He agreed there was no doubting the validity of your discoveries but he was clear in his opinion that the motion of the Earth was another matter. He said that it had in it something of the incredible, especially because the Scripture presents a contrary view. Madama Cristina quoted Psalm 103.'

'O Lord my God, et cetera, et cetera, thou fixed the Earth upon its foundation, not to be moved for ever,' said Galileo.

'And the book of Joshua in which the Sun is commanded to stand still for a whole day in the middle of the sky. I had to turn myself into a theologian and argue against such obvious interpretations. The younger members of the family came to my assistance, but nothing would sway the Duchess.'

'And Boscaglia?'

'He said nothing more, just looked pleased with himself.' Castelli took a deep gulp of his drink. 'Can we really be spreading heresy?'

'Don't be absurd. Both Holy Scripture and nature are emanations from the divine. In the case of the former, it was dictated by the Holy Spirit. In the latter, it was emplaced for us to see. The two cannot be at variance.'

'But, as much as I believe that, I can still see the contradictions.'

'Remember this: while the Scripture cannot be in error, those who expound its meaning are only human. They can err, especially when they base themselves on simplistic literal meanings. Let's find the exact passage and put our minds at rest.'

Galileo picked his way to the bookcases lining the walls. He pulled a well-thumbed copy of the Bible from the shelf and flicked through it, arriving quickly at the book of Joshua. He scanned the pages.

'Remember, the Bible tells us how to go to Heaven, not how Heaven goes. Here it is: "the Sun halted in the middle of the sky".' He looked at Castelli. 'Seems to me that this is exactly where Copernicus tells us the Sun is to be found: in the middle of everything. I would say that the Copernican system makes more sense of this passage than Aristotle's system.'

Castelli looked troubled. 'I wish I had the confidence of your delivery. How I would love to explain this to the Grand Duchess.'

'Then I will detail these arguments in a letter that you will take back to Pisa. Show it to the Grand Duchess – and anyone else who asks. The time for doubts is over. I will demonstrate once and for all that Copernicus is not in conflict with the Bible.'

Castelli's face lifted. 'If it will help spread the true word of God, I will copy it a hundred times and send it to all corners of Christendom.'

24
Prague, Bohemia

Fighting broke out again in the Old Town. In Karlova Street, Barbara sat on the bed with her arms clutching her folded legs. She rested her forehead on her knees and whimpered at every noise.

Kepler perched nearby, attempting to read. He carried a stone of emotion inside him. Each day, it grew heavier.

The raiders had returned under cover of darkness, and seemingly in greater numbers. There was a succession of heavy thumps. The windows rattled in the house, making Barbara cry out.

'Cannons,' said Kepler. He crept to the windows and peered through a crack in the shutters. He glimpsed a column of men in armour striding down the road. These were not irregulars; they were well disciplined and equipped. This must be Matthias's army, arriving at last.

'Papa?'

It was Susanna at the doorway.

'Child, are you feeling better?'

She nodded as if it were a silly question, rubbing the sleep from her eyes. Her face was still pockmarked, but there was more colour in her cheeks. 'Ludwig is crying.'

Frau Bezold was already in the children's room when Kepler arrived. 'I think he's hungry,' said the housekeeper.

'Thank God, they're getting better.'

But then they turned to Friedrich.

'He's burning up,' said Frau Bezold.

The children's pustules had broken several days ago, spilling their clear fluid. While the other two had grown stronger, Friedrich continued to weaken.

She lifted him. The boy's body lolled in her grip.

'I must fetch Doctor Reichard,' said Kepler.

Frau Bezold stood in his way. 'You can't go out just now, it's too dangerous. What if something were to happen to you?'

'What choice do I have?' He pushed past her and fled from the house into the hot night.

He checked the street. Figures were stationed at the mouth of the bridge. This was something new. Kepler peered at the checkpoint, eager to gauge any information he could about their identity.

He lingered too long.

'You there!' one of them called.

Kepler backed away.

'Halt, I say. In the name of Emperor Matthias.'

Emperor Matthias. So Rudolph's brother must already have taken the city. A hundred questions clamoured for his attention, but he shoved them all aside and bolted for the alleyways.

At the doctor's house, he rapped smartly on the door. The doctor looked in need of physic himself. He wore a sweat-stained shirt and smoothed the few strands of his hair with grubby fingers. 'Johannes, if this is about your wife . . .'

'It's my son, Doctor Reichard. He needs your help.'

'Dysentery? It's breaking out across the city.'

'No, smallpox and fever.'

The doctor looked over Kepler's shoulder, surveying the street. 'Is it safe?'

'Yes,' Kepler lied.

'I'll fetch my bag.'

An awful sound filled the house as Kepler returned, the panting Reichard a dozen steps behind. They found Barbara lying on the floor of the bedroom, tearing at the floorboards and howling. In the corner of the room, Susanna clung to Frau Bezold, head buried in the housekeeper's bosom, neither of them daring to approach Barbara's stricken form.

For a moment Kepler thought that Barbara was in the throes of another seizure, but the movements were not the same. Then he noticed Friedrich, and the reality of the situation became terribly clear.

On the bed, his son lay completely still.

Barbara's wracking sobs somehow helped him keep his own composure. He kneeled slowly and ran a hand over her back. 'Barbara,' he said. His voice sounded stern, though that had not been his intention.

She lashed out, arms flailing, catching him with a ragged fingernail across the eyebrows. Instinctively he raised his hand to strike back but at the last moment plunged it downward to grab her hand.

'Leave me alone. Look what you've let happen,' she wailed.

He squeezed her wrists tightly. They ended up staring at each other, breathing fast and hard, tears streaming from their eyes. 'You're hurting me,' she said.

Kepler threw her hands away, jumped to his feet. Susanna was crying silently, still entwined around the housekeeper.

The doctor knelt beside Barbara. 'Drink this, Frau Kepler, it will help you.' He touched a measuring glass to her dry lips. Initially she gagged on the preparation, but then seemed eager for the potent mix. Within minutes, her breathing slowed and she sat back against the bedpost, eyelids drooping.

'Let's get you into bed, madam,' Frau Bezold said calmly. She stroked Susanna's head and untangled the frightened child, then guided Barbara out of the children's room.

'I will arrange for an undertaker to call,' said the doctor, 'if they're not already overrun.'

Kepler nodded dumbly.

Frau Bezold took to praying every day in the children's room. Susanna would cry while clutching Astrid for comfort. Kepler would stand outside and listen, unwilling to step inside in case his heart truly broke. Only Ludwig was untouched by the tragedy.

Barbara refused to get up, looked blankly at her food and would not talk. She stared at the ceiling or at the window where all she could see of the wounded city were the roofs of the houses opposite.

One day Kepler heard Barbara's food tray and its contents crash to the floor.

'I give up with her,' said Frau Bezold, passing him on the stairs.

In the bedroom he cleared up the mess. 'Barbara, you must eat. How long can you last like this?'

She did not answer.

He would sit often on her bed, talking about anything that came into his head. Mostly it was about his work, sometimes about their children. She rarely acknowledged his presence.

'I saw Jan today at the university. He tells me that Rudolph is still alive but confined to his Chamber of Arts. Matthias has taken over all the running of the Empire.' There was no news of von Wackenfels. He had disappeared. 'Let us hope my blond friend escaped in time,' said Kepler. Anything else was unthinkable.

The children spent time with their mother daily, but her eyes had stopped following them around the room. Everyone knew that Friedrich had been her favourite; Kepler's too, if he were being honest. The boy had moved like her, spoken like her. His easy nature and quick smile were like the Barbara Kepler had fallen for all those years ago. The Barbara he still searched for in her eyes. But they were faded and distant, their gaze locked on something unfathomable. Occasionally they looked at her battered prayer book, but even that ritual was on the wane. As the days passed, her face began to sag and her skin took on a grey pallor.

Frau Bezold looked after the children, becoming more of a grandmother than a housekeeper. One night she knocked gingerly on Kepler's study door. She was sniffling as she explained that she could no longer cope. One of her last remaining teeth had simply fallen from its socket, too tired to hold on any longer. She felt the same way.

'I know,' said Kepler.

The next day, Kepler sat on his wife's bed. 'There is nothing to keep me in Prague any more. I've been thinking that perhaps we should move on again. I thought somewhere more like Graz, back in Austria where you will feel more at home.' He watched her, willing some kind of reaction. 'There's a District Mathematician needed in Linz. We could go there; somewhere with no memories, somewhere we can start again.'

Frau Bezold entered the room carrying a pile of laundry and took off the top sheet, flapping it over the bed.

Barbara spoke, her voice so weak it was barely recognisable. 'Is this the cloth of redemption?'

Kepler felt his eyes dampen with tears. 'You have no need for that.'

That night, Barbara's soul departed. She released it so modestly: one moment she was alive, the next moment her jaw dropped and Kepler knew she was gone. Her body remained upright in bed, as unmoving as it had been in her last few weeks of life. He did not call out to her or shake her or make any attempt to revive her because he knew that all she had wished for was to see Friedrich again. This was the reason for her death. How could he deny her this greatest wish?

Her soul had been so deeply wounded by the death of her precious boy that she could find no reason to continue living on this bitter Earth. Despairing of a future without Friedrich, she had achieved her final journey as easily as they could have walked across one country's border with another. He almost admired her resolve, perhaps even envied her a little. *She is with God now, all is well*, he told himself. Yet inwardly he felt as though she had set him on fire. He had loved her with all his heart but it had not been enough for her. She had left him. He could find no words sufficient to express the pain.

On the morning of her burial, Kepler pushed her rigid body on a tumbril through the muddy streets. The act felt as insensitive as the wooden cart itself, clunking through the ruts. Detached and downcast, he trailed his children and the few friends brave enough to be seen at a Lutheran funeral. Frau Bezold, clutching self-consciously at her crucifix, had bade them farewell at the kitchen door with a plaintive 'I cannot' at his request for her to join the procession.

The soldiers they encountered stood aside, though Kepler noticed the captain watching carefully, scanning each face and making sure the swaddled body was not some kind of decoy. He scowled at the soldiers, willing them to challenge him but none met his gaze. Robbed of the confrontation he had been expecting, perhaps even hoping for, Kepler trudged on.

At the graveside, it began to drizzle. Barbara was laid upon a set of straps, ready to be lowered into the dampening earth. The fine rain hid Kepler's tears as he insisted on preaching his sermon. His voice sounded flat and unemotional even to him.

'Some things can only be truly appreciated in retrospect. The movement of the great circles of Heaven can be reckoned only after they have been watched to turn an entire revolution. So it is with human lives. Only at the culmination of a lifetime can we strip away our own feelings, our own hopes and fears for what they may achieve or do. In death can we see them in their entirety and finally understand who they were and why they were so special. Barbara was special to me in every way . . .'

So why did she abandon me?

'She was a person who . . . She . . . She . . .'

The words stopped. He glared at the body. He imagined screaming at her for causing him so much torment. Why make such a selfish choice to be with Friedrich instead of with him and Susanna and Ludwig? Shaking in his rage, he forced his eyes away from the lifeless bundle and scanned the straggle of humans watching him. There were his two dear children, upset and puzzled.

Brought to his senses by their scared faces, he found himself talking again, though where his words were coming from he could not say. He was the mouthpiece of some unknown power. 'Though it is hard to see it at the moment, there must be harmony in this world. God's perfection cannot allow it to be otherwise. The terror of the recent past reminds us that the harmony must be so grand that it reduces all human woe to triviality, as the sound of a buzzing fly is surpassed by the mighty crash of the ocean. Now at peace, Friedrich and Barbara are free to not only hear this great harmony but to rejoice in it. For those of us left behind, we can only wonder and wait until it is our time to hear it, too.'

Kepler felt a tremor of recognition pass through him, substantial in meaning yet incorporeal in form. The first time he had felt this had been in the schoolroom in Graz, when he received his vision of the planetary arrangement. Now it had happened again. For the second time in his life, Kepler knew that God had visited him.

The city was calm at last, but everything was different. Kepler had been confirmed as Matthias's mathematician, but he was not required at court. The new Emperor preferred to take advice from his generals rather than his stargazer. Kepler was grateful for that; no more time wasted casting horoscopes for things they could not be applied to anyway.

To fill his days Kepler had buried himself in Tycho's ledgers, looking for the harmony he had so clearly, if briefly, perceived at Barbara's graveside – her parting gift to him before she took her place with Friedrich in Heaven. He told himself that he would eventually return to the composition of *The Rudolphine Tables*; his oath to Tycho alone meant that he could not forget them, but for now he had to follow his heart and look for the harmony. It was the only way to make sense of all the suffering, and there was no one to stop him. Rudolph had been deposed and Tengnagel was missing, presumed dead in the fighting.

Shut into his study in Karlova Street, he could temporarily forget about Barbara and Friedrich. But every time he left the cramped workspace he had the momentary thought *Where's Barbara?* followed by the painful realisation. During the evenings, he would stare at her empty chair by the fire and her prayer book on the table until he slipped into a fitful sleep in his own chair, or he would drift to the children's bedroom and sit on Friedrich's cold bed.

Barbara's grief at the loss of their son had been so intense that his own had been eclipsed. In the emptiness of the night, the agony of this untended wound tore at him. He looked at his surviving children, their small chests rising and falling in sleep, and wished he could find solace in them. He loved them more than ever but all he could really feel were his losses. Neither was he equipped for their constant demands. Frau Bezold did her best with them but she was old and no substitute for a mother.

Letters of condolence from friends and acquaintances trickled in from across the Empire as the news spread, and it was in one of these that Kepler saw the solution. A friend that Barbara had made during her first few weeks in Prague was now living in Kunstadt having been widowed. He remembered her well; she had been tall and youthful with a ready smile. Her letter offered help with the children.

So, why stay in Prague when everything reminded him of what he had lost? He had written to her asking whether she could look after the children while he went on to Linz to claim the job as District Mathematician, teaching in a small school in a Lutheran stronghold, and she had readily agreed.

Kepler left the house two months after Barbara's funeral with his son and daughter and all the possessions he cared for. As the three of them huddled together on the wagon, the clouds broke. Hunched beneath their cloaks and hats, Ludwig started crying and Susanna did her best to distract him by dancing Astrid on his knees. The rain cascaded from the high roofs to gather on the earth and run in rivulets down the streets. Frau Bezold waved them goodbye and shrank into the background as the wagon rolled off into the cold greyness. Soon, she was lost to the crowds, and Kepler stopped looking back.

The city's beleaguered inhabitants were doing their best to go about their business but waiting in line at the various checkpoints made everything slow and tense. After several anxious moments at the city gate, the family left Prague behind. Some way beyond the city walls, they came across a column of Matthias's troops, heading back in. They were returning from a mopping-up operation in the surrounding countryside. In their midst was a column of prisoners, shuffling barefoot in the mud. Red welts showed at their ankles and wrists where the shackles chafed, and chains swung between each man, draped from one iron collar to the next.

Kepler nearly exclaimed, for there among them was Tengnagel, looking as miserable and bedraggled as the others. Their eyes locked. There was a strange expression of anguish on the captive's face as he stared at Kepler and the wagon of possessions behind the family.

Peeping out from between the waterproof coverings were the tattered corners of Tycho Brahe's astronomical ledgers.

25
Rome, Papal States

Cardinal Robert Bellarmine tutted loudly. It was the nearest he had come to anger in a long time. For a moment, he thought he might succumb to temper, an aspect of his personality he thought he had left behind, along with his youth, at some point during the last seventy years.

He put his usual equanimity down to having seen so much that it was impossible to be surprised any more; or simply that he could no longer muster the energy. Whichever was the true explanation, he did not care at the moment. He suppressed the angry impulse and spoke. 'Most people become wiser with age, but I swear you become more impudent.'

'But you must resent him a little,' said Cardinal Pippe.

'I am his loyal servant, how could I resent His Holiness?'

'Because he was appointed instead of you. Until the last minute, everyone thought you were going to be Pontiff.'

'That was eleven years ago. How do you dream up these fantasies?'

They were sitting in Bellarmine's office. Pippe sprawled in an upholstered chair in front of the desk. 'You must look at him and say to yourself, "That could have been me." '

Bellarmine narrowed his eyes at the younger cardinal. 'No, I do not. This conversation is closed. We have far more serious matters to consider today.'

Bellarmine returned his attention to a letter that had been forwarded to the Office of the Inquisition. There were seventy cardinals in Rome, all serving the Pope as advisors and preaching to the city's inhabitants. As inquisitors, Bellarmine and Pippe had the additional role of combating the spread of

Protestantism and, to do that, they needed to be constantly alert for those seeds of doubt that could grow into heresy.

This letter, as with all the serious cases, ended up on Bellarmine's desk. He picked up a hand lens and read through the contents once more. There was no doubt about it; this particular seed had put down roots and needed weeding out.

The silence was broken by a knock on the door.

Bellarmine glanced at Pippe, who took it as his signal to open the door. He greeted the new arrival. 'My aching legs thank you for making the journey across the city today.'

'The least I could do,' said Father Grienberger.

'My condolences on the passing of Father Clavius. He will be greatly missed.'

'You are very kind, Cardinal Bellarmine. He was a mentor to us all.'

'Indeed. He was a loyal servant to the Inquisition Office. And my congratulations on your promotion to Professor of Mathematics.'

'Thank you. I hope to be a loyal servant to your office, too.'

Bellarmine spread his bony fingers across the letter. 'This is suspected of containing heresy, and I have been charged with its investigation.'

'Who wrote it?'

'Galileo.'

Grienberger's expression wavered, sparking Bellarmine's curiosity.

'He claims that his discoveries have proven Copernicus. And he is playing at theology to justify a Sun-centred universe,' said Bellarmine.

Grienberger tapped a finger to his mouth before speaking in an undertone. 'Galileo must be stopped. We cannot risk him provoking a papal decision at this delicate time.'

Bellarmine became impatient. 'Why not? Wouldn't it be better just to bring an end to this talk, once and for all? I thought we had an agreement: Galileo would stick to describing his discoveries as facts but attempt no justification or interpretation.'

Grienberger glanced at Pippe, who was sitting with his arms folded looking stern, and back at Bellarmine. 'May I speak in confidence?'

Bellarmine concealed his annoyance with a nod.

Grienberger spoke without lifting his gaze, as if addressing the leather on Bellarmine's desk. 'Galileo now has clear evidence that Aristotle's arrangement of the planets cannot be correct.'

Bellarmine pinched the bridge of his nose. 'Explain, please.'

'Some time ago he sent word of a discovery to Prague. He coded it in the form of an anagram. *Haec immatura a me iam frustra leguntuory . . .*'

'These immature things I am searching for now in vain,' Pippe translated. 'It makes no sense.'

'. . . the solution is: Cynthiae fuguras aemulatur mater amorum.'

'The mother of love emulates the shape of Cynthia. I am none the wiser,' said Bellarmine.

'Galileo alludes to the fact that Venus displays the same crescent to gibbous phases as our Moon. We, too, have charted this with our telescopes at the college.'

'What does it prove?'

'As well as changing phases, the planet grows bigger and smaller as it approaches and then recedes from the Earth. The way it does these two things together can only be explained if Venus moves around the Sun.'

Pippe gasped.

'We are sure of our observations; there is no mistake. Venus orbits the Sun, not the Earth.' Grienberger sounded grave.

'And Galileo knows this?' asked Pippe.

Grienberger nodded slowly.

'Then he is dangerous,' said Pippe, rising from his seat.

Bellarmine steepled his fingers. 'When were you going to tell us of these things?'

'With a matter of such magnitude we preferred to collect more observations of the other planets before approaching the theologians. We think Mercury behaves in the same way, but we need more observations to confirm this. If so, both are clearly in orbit around the Sun.'

'You no longer have the luxury of time.' Bellarmine lifted the letter.

Grienberger spoke quickly. 'It would be better if this matter were conducted quietly. We have representatives in Florence who could talk to Galileo, reason with him.'

'You're too late. Galileo has returned.'

'To Rome?'

'Trying to gain another audience with His Holiness, to justify his letter.'

'That cannot happen.'

'Oh, we will see to that,' said Bellarmine, 'but from what you say, you actually favour a rethinking of astronomy.'

Pippe threw his arms in the air. 'On the say-so of this Galileo?'

'On the say-so of the Roman College,' Grienberger corrected. 'We have our own telescopes. We have made our own studies. There is no doubting these observations, but if His Holiness is provoked into a hasty ruling against Galileo, it could set us back decades – perhaps centuries.'

Pippe stared at Grienberger. 'We cannot go rearranging the heavens. God placed the orbs just as he placed each and every one of us in our correct stations. After our lifetime of faithful service, we receive our reward in Heaven. If we start rearranging the planets, what's to stop people rearranging their lives? No one will know what to believe. There will be mass panic. Society will break down. What will prevent the peasants demanding land or riches? They could reject our authority altogether.'

'Calm yourself, Cardinal Pippe.' As Bellarmine spoke, he felt a restless urgency. *Panic* – another sensation Bellarmine had thought banished to his youth. The universe was coming apart around them, and not even the Jesuits knew how to stitch it back together.

Pippe looked at him pleadingly. 'Even if these observations are correct, we must suppress them. There's no sin in concealing a truth if it serves a higher purpose. The simple folk will not know how to interpret this.'

Bellarmine hoped he appeared more confident than he felt. 'Gentlemen, we have a lot of work to do and not much time to do it. Here is what I propose . . .'

Galileo crunched across the gravel outside the Tuscan ambassador's residence. His arthritis was back, gnawing at his joints, and the white of his beard had crept to his head since he'd last stayed in Rome. He defied both reminders of age, forcing himself to move with the speed of a younger man.

Clouds of impatience gathered inside him as he waited. As time stretched on, they became droplets of doubt that his visitor would keep the appointment. Galileo circumnavigated the knee-high hedges of the formal gardens, wondering what he would do if the young cardinal did renege on his promise.

A footman stood in the doorway, supposedly to attend to Galileo's needs but really to keep an eye on him. Since his arrival this time, Galileo had been regarded with suspicion. The ambassador made his reluctance to offer hospitality perfectly clear and lectured the astronomer on the importance of keeping a low profile.

Galileo dismissed it all as the product of a timid mind and considered having a quiet word with the Grand Duke when he returned to Florence. The ambassador was clearly the wrong man for the job.

This morning, in preparation for his visitor, Galileo had read through his new manuscript once more. Before leaving his apartment, he had wrapped the papers around his left forearm and slipped on his jacket. Now, arms folded, he held the precious document in place.

He stayed close to the fountain. The splash of the water would help mask their conversation – if the cardinal ever appeared.

A flap of robes emerging from the house drew his attention.

At last, thought Galileo, straightening himself.

The cardinal hurried down the villa steps and across the courtyard. What little was visible beyond the yards of red silk betrayed his youth. His cherubic features were as yet unblemished by the ability to grow a full beard. He was quite out of breath when he stopped in front of Galileo and clearly embarrassed.

'I am new to Rome. I lost my way.' His voice was so thin that Galileo strained to hear it over the gurgling of the fountain.

The footman watched from the doorway.

'It is of no matter . . . my name is Galileo Galilei.'

'I know who you are, signor. I am Alessandro Orsini and I am humbled to be in your presence. I have read your books and letters. You are a great philosopher.'

Galileo felt the tension ebb from his lower back. 'I am pleased that you have understood what too many so-called philosophers have failed to recognise. If my work has taught me anything, it is that it takes a certain youthfulness of mind to appreciate my ideas. They are revolutionary.'

'Some say dangerous.'

'The Holy Office?'

Orsini nodded.

'You see, this is how I am treated; gagged, while my enemies are given free rein to speak against me.'

'A letter of yours aroused their suspicions. You are under investigation.'

'And that is why we have to act decisively. I believe that my enemies have introduced errors into the copies, designed to incriminate me. I have written a fuller account, more detailed, which includes a discussion of the tides. This proves the Earth moves through space. If it could be shown to His Holiness, I believe the matter could be brought to a swift conclusion.'

The cardinal swallowed. 'You wish me to show something to His Holiness?'

'We must act now before these ideas are banned by those who would block progress. It is people like you who will have to live with the injustice if the closed minds win.'

Orsini tugged his ear nervously. 'I shouldn't even be talking with you.'

'I wish only to clear my name. You say you have not long been in Rome – do you know the Campo dei Fiori?'

'The piazza, yes.'

'And you know the name Giordano Bruno.'

'The heretic.'

'Then you know what happened to him in the piazza.'

'Burned alive,' Orsini whispered.

'Do I deserve that same fate?'

Orsini looked at Galileo, eyes wide.

'It is a matter of life and death, Cardinal.'

Orsini was breathing deeply. 'If I show this to His Holiness . . .'

'You will be doing the Catholic world a great service, Alexander.'

'It's Alessandro, signor.'

'A name that will doubtless be remembered for its part in bringing God's truth to the wider Catholic world. This will make your name in Rome. Will you do it?'

The young man thought long and hard. He swallowed again and looked up. 'Yes, in the name of truth, I will.'

Galileo turned his back to the footman at the villa and slid the manuscript from his sleeve.

There was a harsh rap on Bellarmine's office door. The Swiss Guards burst in before he could answer. The quartet's leader informed him that he was summoned at once to see Paul V and they marched him off, sandwiched between their striped uniforms. He felt as though he himself were under arrest.

The sun shafted through the windows in horizontal beams of orange light. Often at this time of day, Bellarmine would watch the view and glory in the softness of the colour. He would allow the beauty of Rome to warm him. Today he could pay it scant attention. His knees throbbed.

'Gentlemen, may we travel a little more slowly? I am an old man.'

The soldiers displayed no emotion but reduced their pace.

The Pope's chamber was similarly wreathed in orange. A young cardinal stood to one side with his head cast down, flanked by guards. He clutched a sheaf of papers. Bellarmine struggled to recall his name. *Orsini? Yes, that was it, Alessandro Orsini.*

The Pontiff barely waited for Bellarmine to kiss the papal ring. 'Cardinal Bellarmine . . .'

'Your Holiness, I came as soon as I could.'

'Our gullible young friend here has a new document from Galileo . . .'

Bellarmine felt his stomach drop away.

'. . . My patience with our Florentine astronomer is growing thin. The first I heard of him was that he had discovered a few twinkling lights. Now *all* I hear is that he wants to rewrite our beliefs to incorporate the work of some fellow called Copernicus – and he sends my own cardinals to me. I want an end to this nonsense.' He snapped his wrist in a beckoning motion to Orsini. The young man scuttled over, holding out the manuscript. The Pope flicked his head towards Bellarmine, making Orsini proffer the papers to the old cardinal.

Bellarmine accepted them with a peculiar sense of inevitability. Thinking back to Galileo's first visit to the Vatican, he realised that something had told him this day would come. For all his charisma, Galileo was as naive as a schoolboy about the workings of power and influence inside the Vatican. Bellarmine had spent his whole life navigating these choppy waters and was still not certain he always understood the way the wind blew.

Of course, years of pious work could build trust that gradually transformed into respect, so how an astronomer from Padua thought he could simply jump in and set the agenda was beyond belief. Galileo had been granted an audience with the Pope and received his praise. That should have been enough for any man. Had Galileo really thought that it gave him the right to appeal directly to the Pope's supreme authority and demand a rewriting of the Bible over a few twinkling lights?

The Pope addressed Bellarmine: 'Convene a tribunal at once. I want to know – and I want to know quickly – are Copernicus and Galileo heretics?'

Bellarmine suppressed a sigh. He looked from the manuscript to the Pope. Galileo had destroyed himself.

26

A complaint to the Holy Office of the Inquisition usually arrived as a rumour, sometimes as a suspicious letter forwarded anonymously. It then became a piece of gossip that passed from inquisitor to inquisitor. Most of the scandals quickly fizzled out; only a few would catch fire and exercise the clerics.

Gossip was a way of sorting the wheat from the chaff, and, in so doing, cases were shaped long before reports needed to be written. It also meant that the inquisitors could spend most of their time going about the more important work of supplying anti-Protestant literature to the Catholic lands.

Bellarmine had been working on such a manuscript when this irritating business with Galileo had erupted, robbing him of the time to hone his arguments to counter Lutheran propaganda. That was the important work, to spread the Truth. Every day he spent investigating Galileo made him resent astronomers – all of them and their petty nit-picking – a little more.

Now, thankfully, the investigation was nearly at an end. At the Pope's request, he had appointed eleven pre-eminent philosophers and theologians to review the case. Their decisions would form the basis of a papal edict on Copernicus. After weeks of consideration, they had reported back. The eleven reports sat in front of him, fanned out on his desk: eleven assessments, eleven condemnations.

He lifted his eyes from the reports to Grienberger, sitting in the chair opposite. 'I asked the inquisitors to judge the work of Copernicus on two counts. Firstly, that the Sun is the centre of the universe and therefore immobile. On this they answered in one voice: that this thinking is in direct contradiction to the

Holy Word and should therefore be considered not only formally heretical in a religious sense but absurd and foolish in a philosophical one.'

Grienberger remained silent.

'Secondly, that the Earth is not the centre of the universe, nor is it immobile, but it moves as a whole and also rotates to give us night and day. On this they also judged the evidence false. I am to draft a papal edict to reflect these views.'

Grienberger pressed his palms together. 'I urge you to be cautious. Although we have no observations that prove Earth is moving, we know that Venus circles the Sun. For that reason alone, we have to find an alternative to Aristotle.'

'So not Copernicus, yet neither Aristotle. We cannot allow this confusion to continue.'

'Neither can we allow an edict that blocks the progress of knowledge.'

'A strong Church with strong leadership is what I believe in. The Jesuits have always stood for that too.'

A desperate look entered Grienberger's eyes. 'Tycho Brahe developed a system in which all the planets orbit the Sun, yet the Sun moves around the Earth. Thus we maintain our place in the centre of things, satisfying the new observations and the theologians. Given the political considerations, we should move towards this interpretation. I do not think that these are in conflict with your edict.'

'Maybe not, yet I have eleven judgements of heresy to contend with.' Bellarmine indicated the papers on his desk, 'There will be those who view any tinkering with astronomy as suspicious. And I believe that you are at heart a Copernican, which could soon make you a heretic, along with Galileo.'

Grienberger spoke quickly. 'The way forward is this: in Copernicus's book is a preface that makes the distinction between talking in mathematical hypotheses and asserting truth. It states that the ideas Copernicus discussed were only to be taken as a hypothetical way of understanding the motion of the planets, not as truth. So, the only thing that you must keep from the edict is the word heresy. If you soften the language, then we can at least proceed with gathering the measurements. But if you stop us, the Lutherans will forge ahead. They will make us look like simpletons.'

The thought struck home with Bellarmine. *Why could this not be simple?* He took some time to consider the suggestion then nodded wearily. 'Very well, I will see what I can do.'

'And what of Galileo?'

'He must be made to stop. I will summon him myself.'

Grienberger stifled a yawn as he opened the tripod of the telescope and shuffled it into the lee of the Roman College's bell tower. He pointed the leather-bound tube towards the east and waited. From up here on the college roof, he had a clear view to the horizon, and this early in the morning the city's silhouette was just beginning to fill with colour.

The black of the sky was giving way to dark blue, but the stars still shone from directly overhead. Grienberger ignored them, determined not to be distracted and miss the vital observation. This morning, he was here to see Mercury. He had to convince himself whether it showed phases like Venus. If it did, perhaps he could persuade Bellarmine to soften the edict's language even further, to acknowledge that reinterpretation was needed. The cardinal meant well but he did not have sufficient knowledge to take these sweeping decisions.

Elusive Mercury never strayed far from the Sun and so could only be glimpsed in the twilight sky before sunrise or just after sunset, but never in the dead of night. Now that Grienberger thought about it, that fact alone should be enough to persuade anyone that the planet orbited the Sun. If it had been free-wheeling about the Earth, it should appear throughout the night sky totally disconnected from the Sun; sometimes far from it, sometimes close by. Yet it hugged the luminous orb as if afraid to venture into the dark; exactly the behaviour to be expected if it were orbiting the Sun.

He waited patiently, scanning the horizon, until . . .

There!

A clear spot of light had just risen above the dawn horizon. Now he had to move fast because the Sun was close behind. He aligned the telescope, having now grown quite expert, and stooped to peer through the tube.

Yes, it is Mercury.

Perceptibly larger than a star, the telescope showed it to be bobbing in the soupy atmosphere. Grienberger's eye chased the tiny blur around the field of view. Several times it jumped out of sight as it rose into the sky, so he patiently lined it all up again. He could see the planet was extended in size, but it was too blurred to make out a shape. He kept staring.

Some moments later, the atmosphere went stock still, as if God had looked down and frozen it solid with a mighty breath. The stillness lasted just a split second, but it was enough. Grienberger saw Mercury was a crescent; its rounded back held downward to the horizon, to where the Sun was about to rise.

In that instant, Grienberger knew that the planet orbited the Sun. He could see the arrangement in his head as clearly as his shadow on the bell tower. A wave of elation flared inside him. He felt his eyes creasing as a smile spread itself across his face, and his arms lifted in triumph. He had a sudden urge to laugh out loud and send his joy echoing across the rooftops. He whirled to face the city and looked straight at a gaunt figure, silhouetted against the blush of sunrise.

A stab of fear mingled with embarrassment inside Grienberger. He dropped to one knee and bowed his head. 'Father General.'

Praepositus Generalis Claudio Acquaviva did not move or say anything immediately. He just contemplated Grienberger before eventually opening his mouth. 'And what makes you happy enough to dance this morning, Father Grienberger?'

'I have seen Mercury for the first time.' The words sounded hollow.

Again there was a long silence. 'I have been speaking with Cardinal Bellarmine. He is concerned about you.'

'There is nothing to be concerned about. I seek only God's natural knowledge.'

There was no pause this time. Acquaviva's throaty voice rose in volume. 'Father Grienberger, this has gone on long enough. When you joined the Society of Jesus, you swore obedience above everything. That includes natural knowledge. You are the Professor of Mathematics now and you are expected to uphold the highest traditions of the Roman Church. We subscribe to the teachings of Aristotle and we are pledged to defend it to our dying breath. *You* are pledged to defend it to your dying breath. Banish all thoughts of rearranging astronomy from your mind and cut yourself free of all those who might drag you down with them. Do I make myself clear?'

Grienberger clasped his hands together and stared at the floor. 'Yes, Father General.'

'I expect to see you at matins today. Afterwards, I think you have quite a lot to confess.'

'Yes, Father General.'

After making the painful journey to the entrance of the Inquisition's Palace, Bellarmine removed his biretta and waited for Galileo. He hoped the stubborn Florentine would recognise both gestures as marks of respect. Although why he should favour the astronomer so was beyond him.

As the carriage approached, Bellarmine felt some pangs of sympathy. He was about to deprive Galileo of the right to speak on a subject the astronomer obviously believed.

Bellarmine tried to picture what it would be like if the situation were reversed. What if his belief system were outlawed? For instance, if the Lutherans were to gain control. His mouth dried at the prospect.

What was he thinking, comparing true religion to mathematical jiggery-pokery? The two were in no way the same.

The carriage door swung open. Galileo stepped to the ground before the vehicle had come to a complete stop. He looked around imperiously.

'Galileo, so good of you to come.'

'Good of you to see me in person.' Galileo's voice was tight.

'Shall we choose a bench in the cloisters? On a day such as this, I would usually love to walk but I'm afraid my bones won't allow it any more.' He held a hand to indicate the way. Galileo strode on, oblivious to Bellarmine's struggle to keep up.

The cloisters were blackened by shadows. Galileo listened as Bellarmine told him the news, scowling at times, looking away at others, seemingly more interested in the strangers walking across the quadrangle. In the end, he regarded Bellarmine with an expression that a father might bestow upon a child caught in a moment of naivety rather than naughtiness. 'You don't appear to understand what's at stake here. While we cover this up, Kepler leaps ahead. We risk this revelation being claimed by the Lutherans, who will cast us as a Church of dullards with no place in the modern world.'

'Wheels are turning to make sure that doesn't happen.'

'The Jesuits.' Galileo said their name derisively.

'They are the arbiters of natural knowledge. They have the real influence and know how to proceed.'

'I have already shown how to proceed.'

'You have no overwhelming evidence for Copernicus and you are no theologian. It is as Solomon wrote, "The Sun riseth and the Sun goeth down, and doth hasten to his place where he arose." Your actions have already set back the Jesuit efforts.'

'But when one is on a boat moving away from the harbour, one can equally imagine that it is the shore receding rather than the boat advancing . . .'

'I reject that argument. I can feel the motion of a boat, I know it is moving,' said Bellarmine.

'You are being fooled by the rocking of the water.'

'I am being fooled by nothing. I see the movement of the boat. My eyes see no movement of the Earth. My body feels no movement of the Earth. Yet I see the movement of the Sun.'

Galileo frowned but said nothing.

'There are those in the Vatican who wanted your thinking to be branded heretical. Had they been successful, you would now be under arrest, and I need hardly remind you of the price for heresy. Is it truly your love of the Catholic Church that urges you to push these ideas, or is it your desire for a place in history? God will judge the worth of your work and whether you shall be remembered. But from now on you are forbidden to hold, defend or teach the Copernican doctrine.'

Galileo rose and walked to the nearby archway. He leaned against the stone pillar, gazing at the brightness on the far side of the courtyard. 'I am gagged and bound.'

'If only you had spoken of the Earth's movement as a hypothesis, not as truth. If you had claimed that it was a mathematical shortcut, then all would have been well. Even Copernicus included a preface to this effect.'

Galileo looked over his shoulder at Bellarmine. 'That preface was inserted without Copernicus's knowledge. He would never have agreed to it.'

'Then it's ironic indeed, for that passage is the only thing that has kept it from being banned altogether. I have placed the book on the suspended list, pending correction. Only after the phrases that talk of the Sun being factually located in the centre of the universe are identified and removed, will the book be allowed back in circulation.'

Galileo blew through his lips.

The action hardened Bellarmine's resolve. 'Attempt any more claims that Copernicus's scheme is the true arrangement of Heaven, Galileo, and you will be prosecuted.'

'I understand.'

After Galileo stalked away, Bellarmine rested against the cloister wall. His chest ached. He took a number of steady breaths. The air did not seem to reach as far down into his lungs as it once did. It took some minutes before he felt well enough to continue the rest of his day. He made his way back to his office with a feeling of despondency, wondering how long it would be before Galileo's name crossed his desk again. As he tried to settle that night, he found himself thinking about someone he had not brought to mind for years: Giordano Bruno.

Don't be a fool, Galileo, he thought. *Don't be a fool.*

PART III
Setting

27

Linz, Upper Austria

1612

Kepler had intended to remain in Kunstadt just a few days, long enough to settle the children with the widow Frau Pauritsch and then to set off to prepare for their new life in Linz. However, in the warmth of her timber house, in a quiet part of town, she had made them comfortable and, apart from the wearying round of condolences, Kepler felt unexpectedly relaxed and content to rest awhile.

He wrote to the school in Linz, who were happy to give him some time. They were thrilled to have the imperial mathematician in their faculty and assured him that the job and accommodation would be available for him whenever he showed up.

The ache of losing Barbara persisted like the rheumatism that gnawed at his muscles during wet weather. In bed at night he would automatically roll towards her side of the bed, seeking her warmth. When he was in the town, he would start if he saw someone of her build and height. Such moments of impossible fantasy that she was still alive were the only true pleasures he could find.

Soon, the sympathy of others had turned to thoughts of matchmaking, and that was when it had all become too much to endure. He knew he needed a wife, if only to look after Susanna and Ludwig, but there could be no love involved, of that he was certain. He could never replace Barbara. Finding a new wife was simply computation, he told himself: she must serve him well and look after the children. Even so, he could summon no enthusiasm for the task.

Then the whole situation had taken an unexpected turn. Frau Pauritsch had appeared in the doorway of his room late one night. Her greying hair was loose over her slender shoulders and she wore only a nightshift. As he shrivelled into the far corner of the bed, she offered to make their temporary arrangement permanent.

He looked at her and tried to convince himself. She would be a good mother to his children; she had treated them with nothing but generosity and kindness since they had arrived; and yet something was wrong. He could not bring himself to respond, and the encounter had dissolved into embarrassed silence.

It did, however, galvanise him to complete his intended journey to Linz. The next morning, after a subdued Frau Pauritsch agreed to continue looking after the children, Kepler walked to the docks. He found a newly constructed barge being loaded for its trip downriver and struck a bargain. His wagonload of possessions was hoisted on board a day later and, with a gentle push from the wharf, his journey to Linz began.

'I'll return for you as soon as the house is ready,' he called to his children on the bank, who waved back dutifully. Frau Pauritsch waved her tear-stained hanky and clutched Ludwig's hand.

He relished the anonymity to be found among the packing cases and wine barrels. Free to savour his memories of Barbara without being made to wallow in them by others, he felt increasingly liberated as the vessel bobbed downstream.

At first he occupied himself by watching the crew. Having no power of its own, the barge was under the command of the Danube. The only control came from a bargee standing at each corner of the vessel, grappling with a rudder. Occasionally one would shout to his colleagues; at all other times the men worked their giant paddles in silent synchrony, reading the river and guiding the craft along the currents.

When he tired of that, Kepler busied himself with numbers, devising a method of calculating the volume of wine barrels – anything to avoid having to compute the best of the possible brides who had been offered.

Linz clung to the riverbank as if afraid to be noticed. Only a few church spires dared to raise themselves much above the wooden quays. Kepler stood on the bobbing deck of the barge and did what he had promised himself he would not do: he thought of Prague. He brought to mind the way towering buildings had overshadowed the Vltava, turning its river into just another element of the city. The Danube was different. It was clearly in charge here, cutting a swathe around the city to hold it in check.

Rain was falling, and the impact of the droplets turned the river into a landscape of splashes. To Kepler, the watery surface looked like a million swarming ants. He stepped out from the barn-like construction that ran the length of the deck and readjusted his hat, creating his own small river that flowed from the brim to his shoulder and then in a splashing waterfall to the deck.

The men pulling on the rudders edged the barge towards the dock, where workers in rain capes waited. The bargees flung coils of rope towards the shore, where their counterparts caught them and heaved the vessel into its berth.

Along the quayside, wooden tripods with ropes and pulleys hung over other barges. Nets bulged with goods and swung in the air as they were manoeuvred from shore to barge or vice versa. Soon Kepler's possessions would be similarly swaying. The space they left behind would be filled with some other paying commodity and the barge would continue downstream, carried by the flow towards the Black Sea.

Once at its journey's end, there was no way of navigating back upstream. The barge would be dismantled, and the wood sold off. The bargees would then start their journey back on foot, following the river paths, each hauling a small raft of goods upstream to make some more money.

Kepler bade his temporary companions farewell as he planted his feet on the ground of Linz, his new home. He found the modest house that the school had organised for him to rent – modest by Prague's standards – and arranged for his possessions to be brought across from the docks. The first things he unpacked were the ledgers. He stacked them in towers around the table in the dining room because there were not enough shelves.

Now was the time to fulfil his promise to Tycho and Rudolph by completing *The Rudolphine Tables*. In the process, he could put his theory of elliptical orbits to their most stringent test yet by using it to compute the positions of all the planets.

He wasted little time in reporting to the school for duty and was surprised to find it smaller than the one where he had taught in Graz. The governor guided him along the corridor to a small but comfortable office.

'What duties did my predecessor perform?' asked Kepler.

The governor looked confused. 'You have no predecessor. We created the post especially for you. It is our honour to host the imperial mathematician as he completes the great *Rudolphine Tables*. We will help you distribute them throughout the length and breadth of the Latin-speaking world.'

Kepler also set about finding the church. It was a simple whitewashed affair with a single tower. Inside, the vaulted ceiling was high enough to be impressive, and the wooden pews gleamed with varnish. Near the altar, a carpenter was putting the finishing touches to a chancel screen. He was smoothing the carved rods with a piece of glass paper, filling the lofty chamber with a soft rasp and the faint odour of wood dust. A swarthy man with a mass of raven hair swept past the craftsman to bustle down the church aisle and introduce himself. 'I am Daniel Hitzler, chief pastor in Linz.'

'Delighted to meet you, Pastor Hitzler.'

'Welcome to Linz and our rapidly growing congregation.'

'I hope to add myself to your number. Johannes Kepler, honoured to meet you.'

'*You* are Johannes Kepler?' Hitzler's voice was laced with disbelief. He squared his stance. 'I must ask you to leave.'

The carpenter glanced in their direction.

'I beg your pardon,' said Kepler.

'You are suspected of heresy.'

It took Kepler a moment to realise that he had heard Hitzler correctly. 'I have suffered for my Lutheran beliefs more than most.'

'You, sir, are at variance with the Lutheran Church. I have been warned about you and your heresies concerning the ubiquity doctrine.'

'Perhaps I can explain my position to you.'

'That will not be necessary, you are not welcome in this church.'

'You have no right to exclude me from communion,' said Kepler.

'I have every right. The only way that I will admit you to my congregation is if you prove your devotion by signing the Formula of Concord.'

'To be Lutheran is to use the mind that God gave you.'

'To be Lutheran is to accept the interpretation of the Bible detailed in the Formula of Concord. Because there is suspicion attached to you, I require that you sign this document before I allow you to take communion here.' Hitzler folded his arms.

In Kunstadt, Kepler had resumed attending services, having been prevented from doing so by the ban in Prague. Despite his early assertions, he realised just how much comfort communion could bring. 'Very well, I will sign the document – with appropriate amendments.'

'Not acceptable,' Hitzler said, raising both hands. 'It is all or nothing. If everyone picks the ripest fruits of faith, how do we maintain our unity? The Protestant Church is divided enough: the Calvinists, the Huguenots – we can have no more splits. Every time we divide, we weaken, meanwhile the Catholic Church grows stronger.'

'Then don't cast me out over one small disagreement.'

Hitzler leaned in, forcing Kepler to take a step backwards. 'Herr Kepler, you carry a plague. If I admit you into my congregation I risk the spread of heresy to all those rubbing shoulders with you. Before I know it, I will lose half my congregation. As a surgeon amputates a limb to save a body, so must you be excised.' The pastor's jaw was set firm.

'I will write to your superiors,' said Kepler. 'You will answer for this insult.'

A number of weeks later, a courier arrived with a letter from Tübingen. Kepler ripped open the seal, eager for the words. He read them in disbelief:

Either give up your errors, your false fantasies, and embrace God's truth in a humble faith, or keep away from all fellowship with us, with our Church, and with our creed.

The words disappeared into a grey fog on the page. Suddenly faint, he stumbled for a chair. Mästlin had finally and absolutely turned against him and, as a result, Tübingen had sided with Hitzler. He had been excommunicated.

That evening, the sunset was a purple bruise spread across the sky, and the night brought people spilling out onto the streets from the taverns and inns. Their attempts to quench their thirst with wine and ale steadily took its toll, and soon the throngs were laughing and shouting. Occasionally they burst into song while all the time drinking more and more.

Kepler pushed his way through the crowds, desperate to escape the silence in the house. Where was Susanna when he needed her, breezing into his study with yet another question? Ludwig, who tested everything to destruction? He yearned for the impossible: Friedrich and the way he used to laugh at everything.

All around him, couples dragged themselves into dark alleys to satiate their alcohol-induced desires and young men scrapped with each other over meaningless points of honour. A listless madrigal rose from the strings of an unseen cittern and Kepler followed the sound to an inn. Inside was a fug of pipe smoke, through which Kepler made out the musician, drifting from table to table, cradling the cittern and coaxing sounds out of it by walking his spidery fingers across the long neck.

Once Kepler had been served, he turned from the bar with a jug of wine and a goblet. There must be a seat somewhere. A man with bloodshot eyes indicated an empty stool nearby. Kepler sat down and sipped his wine. As he did so, the man shook his own goblet. It was empty.

'Why not?' said Kepler and filled it.

His new companion smiled a toothless grin and toasted Kepler, who followed suit. The strange warmth of the alcohol exerted its calming influence. He felt the tension ease from his muscles. He took another mouthful and the tension eased a little more. So he took another mouthful, and another . . .

'Let me get this straight,' slurred his companion. 'You walked away from number three because she was too pretty . . .'

'You're getting confused. Number three was the one who promised to check on my children every day in Kunstadt.' Kepler, too, was having problems pronouncing his words.

'So what was wrong with number three?'

'I met number four.'

'And she's the pretty one?'

'That's right.'

'Which one was the clever one?'

'Number two, offered to me by her mother before I left Prague.'

'Was she pretty?'

'Too young for marriage, if I'm honest.'

'Oh.' There was a pause.

The pair of them leaned shoulder to shoulder on the bench, ignoring the draught from the window behind. Three empty wine jugs sat in front of them.

'Any problem can be solved by collecting enough data and then performing the correct analysis,' explained Kepler. 'From the movement of the planets to the selection of a bride.'

'Mmm. Tell me about the pretty one again.'

'She was young and firm, tall and slim,' said Kepler.

'But not too slim? She still looked like a woman, didn't she?'

Kepler tried to bring his companion into focus. 'Which one of us is telling this story?'

'You are. I'm just trying to help you get to the good bits.'

'Well, don't.'

The man raised his goblet and drained the remains. 'Have you met a rich one yet?'

'No.'

'Is that what you're waiting for?'

'No.'

'You don't want a clever one; you don't want a pretty one; and you don't want a rich one. What *do* you want?'

Kepler reached for the table and pulled himself up. He wobbled on his feet. 'A country girl, with apple cheeks and an earthy laugh.'

'I'd have settled for the very first one I was offered.'

Kepler looked from the stubble on the fellow's chin to the snowstorm of flecks on his dirty shoulders, then stumbled from the bar.

Outside, Kepler found it difficult to locate the correct road. He staggered around the characterless streets hoping for a landmark. Only when he reached the city gates with their dancing torches did things begin to look familiar. There were guards milling around the wooden gateway.

'You sure you want to go out there tonight?' asked the scar-faced leader.

'Just open the gate. I am my own person,' Kepler said indignantly.

The guard shrugged and signalled his men to open the gates.

Beyond the city, the darkness was complete. Some corner of his brain recognised the recklessness of his actions. The surroundings were bound to be crawling with brigands, all willing to slit his throat for the few coins left in his purse. He walked on, convinced that he would feel the icy caress of a blade at any moment.

Topping a small hill he sank to the grass, registering the dew seeping through his breeches but not caring enough to stand up again. He scanned the town. It was a maze of dark streets, broken only by the occasional flame of a welcome torch, or the yellow flicker of candles inside a window.

As he watched, the waning moon lifted itself into the sky, laying its silver light across the rooftops. Kepler estimated that Earth's satellite had six days to go before it would become a new moon again. At that point, it would skulk past the Sun to re-emerge as the thinnest sliver in the evening sky. For a fortnight it would grow to fullness, then progressively shrink again, each phase betraying the changing angle made by the Moon, the Earth and the Sun.

In his intoxicated state, Kepler projected himself to the Moon. From those rugged mountains, the situation would be reversed. Earth would go through phases, growing to fullness and then shrinking away again. As it did so, it would reflect sunlight onto the Moon, easing the passage of the long lunar night.

A day on the Moon would not last twenty-four hours but a month. Each night would take a fortnight to pass with only the Earth to provide illumination. It would hover in the lunar sky like a dragonfly over a summer pond. The lunar people would surely think that it was constructed solely to provide them with light and construct their whole cosmology around it.

But this would be true for the near side only. The far side would endure its long nights in darkness. How desperate would those inhabitants feel? Unloved by God who had determined that they should be born on the far side. Would they wallow in despair?

Perhaps as they looked up at the stars they would realise the truth: that God had cast them for greatness by giving them the gift of astronomy and a fortnight of unbroken darkness every month to exercise their minds.

While the nearside population would revel in their supposed favour from God, those on the far side would make the true advances. They would eventually understand how Heaven worked and, in the process, bring themselves alongside their God.

The light of dawn guided Kepler back down to Linz. The battle-scarred soldier eyed him again as he entered the city alongside the traders and their laden carts. Thankfully the streets had returned to familiarity.

Rounding the final corner before he came to his house, Kepler stopped in his tracks. He must still be drunk; he was seeing things. To his blurry eyes, it looked as if a wizened little gargoyle was huddled on his porch step.

The creature turned its shrunken body and spoke to him. 'Looks like I've arrived just in time.'

'Mother?' said Kepler.

Katharina Kepler busied herself in the kitchen. She fished out clumps of leaves from her travel bag and dropped them on the chopping board. She stoked a fire from the piles of wood in the outhouse.

'You have enough firewood, I see.'

'I am not wanting in that respect,' said Kepler, cradling his pounding head.

She boiled up a variety of the leaves and ladled him a bowl of the steaming concoction.

Kepler wrinkled his nose at the smell. 'Do I drink it or sniff it?'

'Don't be clever. Drink it.'

He took a few sips of the hot liquid. The strong taste was more pleasant than some of the things his mother had made him drink. Before he was halfway down the wooden bowl, his head had cleared.

'What are you doing here, Mother?'

'Is that any way to treat me after I've come all this way? I'm not going to sponge off you. I've brought these so I can pay my way.' She indicated the greenery spread around her. 'Winter's coming up. People will want remedies.'

'You're hiding something.'

Katharina brushed some unseen fleck from her sleeve. The window silhouetted her face, accentuating the sharp outline of her chin and nose. In the years since Kepler had seen her last, the flesh on her face had gone. She had never been a pretty woman but now she was old and ugly. She said, 'There's some business back home I want to avoid for a while.'

'What business, Mother?'

She did not reply.

'What have you got yourself involved in?'

'It's not my fault, it's that Ursula Reinbold. We had an argument a few weeks ago. When I thought it was all over, I invited her in for a drink – gave her my best wine, too. Ungrateful woman said it was bitter and soon after she started saying other things, too.'

'What things?'

'Told anyone who would listen that I gave her a potion that made her sick.'

Kepler's headache returned. 'How far has this gone?'

Katharina sniffed. 'The magistrate summoned me. I thought he wanted to clear everything up, but when I got there, Ursula and her brother were there as well. They'd all been drinking. They called me a witch and demanded that I remove whatever curse I had placed on Ursula. When I told them I had done nothing, her brother drew his sword and told me he would run me through the heart if I didn't reverse the witchcraft.'

'What did you do?'

'I could hardly get the words out, I was shaking so much. I could feel the blade pressing on me. Told them I had nothing to do with any witchcraft. Then he grabbed Ursula by the arm, threw her at me and screamed that she should curse me in return. That's when the magistrate saw sense and called an end to it.'

'Oh, Mother, I'm so sorry. Tell me, what's the magistrate's name?'

'Einhorn, Luther Einhorn.'

'I will write to Tübingen at once. The law faculty must know of these irregularities. Then we must return to Leonberg so that we can bring a case of slander against the Reinbolds.'

'Go back?'

'Not immediately, but it's the only way if we're to clear your name.'

Katharina wiped her eyes. 'I knew you'd know what to do. I've told them all about you. How clever you are, how proud I am of you.'

She placed herself into his embrace, and he closed his arms around her. There was nothing of her, just bones and skin.

A new thought jolted into his mind. 'I'm late.'

He had accepted a request from Baroness Starhemberg to favour her with a reading. As much as he still hated such tasks, it would perhaps be his easiest route into Linz's more rarefied strata. He collected his things, Katharina dogging his every step.

'You can't go out in those clothes. They're filthy,' she said.

'I'm a little behind on my washing.'

His mother's expression was a mixture of amusement and anger, the way she used to look when he was a child and had forgotten to do some chore or another. 'I'll do them while you're out,' she said, 'and I'll cook something, too. We need to put some flesh on you if we're ever going to find you another wife.'

Kepler regretted falling behind with his washing the moment he saw the Baroness. She was wearing an immaculate chocolate brown dress with matching stays. Her eyebrows rose as she took in Kepler's rather shabby appearance.

His mother had flicked some herbal scent all over him as he had left and he had brushed much of the dried mud from his breeches as he had dashed through the streets, but nothing could compensate for the fact that his clothes were well and truly dirty.

'Forgive me, Baroness, I spent the night on the hillside, engaged in lunar observations. I regret my attire is entirely inadequate.'

His remorse was absolute when the Baroness moved to reveal a beautiful young woman a few paces behind her. Dressed in fawn, with the smallest waist Kepler could remember seeing on a grown woman, she was waiting with perfect decorum, the whisper of a smile playing on her lips.

'This is Susanna Reuttinger, my companion,' said the Baroness.

Susanna . . .

'Such a pretty name,' said Kepler. 'My daughter shares it with you.'

Her complexion was as pure as moonlight, and, despite her slender figure, her cheeks were as round as apples when lifted in a smile. Kepler couldn't help but return her smile.

28
Florence, Tuscany

There was little room for manoeuvre in the gloomy loft space. Bent double and silently cursing the awkwardness of his task, Galileo inched past the telescope and its tripod in order to get to the south-facing eaves. He was carrying a knife.

The sound of puffing drew his attention.

'There's no point in you trying to join me, Benedetto. You won't fit through the hatch.' It had been a squeeze for Galileo, but he was not going to admit it.

The monk's balding head appeared through the opening. 'Nevertheless, I'll be able to see the image you project.'

Galileo dug the blade through the plaster and cut a hole the size of a coin. Then, he levered the blade under the exposed roof and eased a tile to one side. A shaft of sunlight split the air.

'Perfect,' he said. 'Now, let us see what this Apelles is talking about.'

Galileo ran his hands across the broken plaster and clapped them to release the dust. The tiny flecks twinkled in the needle of sunlight. Galileo matched the angle of the telescope to the brilliant shaft then manouevred the telescope to the hole, so the sunlight fell into the tube.

Castelli coughed in the plaster dust, drawing the folds of his habit over his mouth. 'Who do you think Apelles is?'

'It's clearly a pseudonym. Whoever it is attacks me and my views, and then gets some German amateur I've never heard of to forward the letters to me.' Stepping over Castelli, Galileo pulled an easel with a canvas into place. The image of the Sun appeared on the surface. It was as bright and as big as Castelli's head. 'There! There they are: the spots on the Sun,' announced Galileo, crouching level with the reflection.

The burning disc was almost featureless except for two spots. They were little more than the size of tack heads and slightly misshapen. Although their centres were as black as pitch, the edges were ragged grey, dissolving into the bright surface.

Castelli had twisted round to look for himself. 'It's just as Apelles says.'

'No, it isn't. He claims that sunspots are the silhouettes of previously unknown moons. But look at them, these are clearly surface markings.'

'So why does this Apelles claim that they're not?'

'It's yet another desperate effort to shore up the Aristotelian doctrine that the heavens are unchanging. If the markings were silhouettes, it would mean the solar surface remains blemish free, and Aristotle lives on. I tell you, I can scarcely believe that such ignorance persists among us. Trying to explain a new phenomenon with old philosophy is a path to ridicule.'

The bell of San Matteo struck. Galileo's mouth opened in alarm. 'You've made me late. Move, Castelli.' Galileo dropped onto his buttocks, sending dust billowing, swung his legs through the hatch and waved a foot in search of the stepladder. Castelli caught his ankle and guided it to the top rung.

'Late for what?' asked the monk as Galileo completed his unsteady descent.

'Virginia takes her vows today. Come, we must hurry.' Galileo wiped his hands across his chest, leaving streaks on the smock. 'Fetch my tunic.'

Castelli looked around, spotting a black garment on a chair. 'This one?'

'Yes, yes.' Galileo held his arms out for Castelli to slip the long piece of clothing up onto his shoulders. It had no sleeves but hung to the floor. Galileo arranged the front to cover the dirty marks. 'Now, let's make haste.'

Despite the bright sky, the chill of winter was in the air. Galileo shivered. 'I don't suppose you feel the cold, do you?'

'Feel it? I love it. I welcome the coldness with all my heart,' said his companion.

The mournful tolling guided them down the narrow street towards the bottom of the hill.

'The question remains, who is Apelles?' puffed Castelli as they entered the shadow of the convent's high wall.

'Oh, didn't I say? Christopher Scheiner of Ingolstadt.'

'How do you know?'

'I compared the handwriting with my collection of letters. Scheiner's fits perfectly.' He walked past the convent's entrance and continued to the church beyond.

'Scheiner is a Jesuit.'

Galileo threw a grin over his shoulder. 'Intriguing, isn't it?'

A familiar figure stood, back turned, by the church entrance.

Galileo came to an abrupt stop. 'Marina,' he said, smelling her familiar perfume. 'I didn't think you would make the journey.'

She looked up at him with puffy eyes. 'You told me you would never let it come to this. You promised me you would find them husbands.'

Castelli dropped his head and shuffled in embarrassment. 'I'll see you inside, Master.'

Galileo paused as the monk moved away. 'Marina, I did what I could . . .'

'Did you? Honestly?'

'There were constraints on me.'

'What constraints? You have the Grand Duke's ear.'

Galileo took in a breath. 'Precisely because I am a member of the Grand Duke's court I have to consider appearances now. Virginia and Livia would need to marry men of standing, otherwise it would reflect badly on me – and on the Grand Duke. The trouble is . . . the trouble is, I cannot raise the dowries to attract such men.' He squirmed at the look of suspicion in her eyes.

'You have money, Galileo. There must be someone.'

'No, there isn't, Marina. It is better that the girls become nuns than marry beneath themselves.'

'Don't lie to me.'

She always could see through him. He looked around for inspiration, some way to break the real news, but there was no way to soften this admission. She had caught him out. He lowered his voice. 'I can find no spouses of standing for them because they are illegitimate. It is not the dowries that is lacking; it is their heritage.'

'You're telling me that if you had married me, our daughters would be married now. I begged you to marry me, and you told me it would never matter.' Tears pooled in her eyes. Galileo reached out to comfort her but the venom in her voice made him withdraw his arm. 'You told me it would never matter! You've treated me no better than your own private whore.'

'I love you.' The words were out of his mouth before he knew it.

She froze at the admission. There was a timeless instant in which Galileo's thoughts were as still as her expression, and then a sharp pain burst across his face. The flat of her palm flashed by his eyes, and his head turned with the force of the slap.

She folded her arms, the way she used to when Galileo had gone too far, and bit her bottom lip. 'You have ruined their lives.'

'I have plans for Vincenzio,' said Galileo hopefully. 'The Grand Duke has promised to make him a Florentine citizen when he comes of age. He will have his pick of the women here.'

'So, I will lose my son to Florence, too.'

'It is Vincenzio's chance for a better life.'

Marina did not look back at him. 'Of course.'

Inside, Galileo took a pew near the front. Castelli edged in next to him. The congregation stood to sing the first hymn. The massed voices of the nuns could clearly be heard through a black metal grille in the wall behind the altar but they were nowhere to be seen. Only afterwards would he be permitted to see Virginia, in the partitioned meeting parlour. Galileo kept watch on Marina's small form, isolated and alone on the other side of the chapel where she had deliberately stationed herself.

After the mass, the Abbess called Virginia's name and addressed her for the congregation to hear, drawing out the important words for emphasis: 'You appear today in your new clothes, the Franciscan habit and veil. Entering into your marriage with Jesus is a profound change in your life. To symbolise this choice, you will leave behind the name you were given and choose a new name. What is the name that God has chosen for you?'

Virginia's clear voice rang through the chapel. 'Henceforth I shall be known as Sister Maria Celeste.'

Across the line of pews, Marina wept openly.

Galileo took his seat in the convent's meeting parlour, a room divided into two by a wall. There was no adjoining door, just a trio of small windows with metal grilles. The room was a perfect mirror of the Franciscan nuns, the Poor Claires as the townsfolk called them. In common with all who took the veil, the nuns had sworn their vows of poverty, chastity and obedience, but, in the case of the Claires, poverty was the defining trait.

Maria Celeste sat behind the grille, her oval face framed by a flowing wimple. She was taller than her mother, with a masculine nose and Galileo's full lips.

She should have been a son.

'Your mother's upset with me,' he said in hushed tones so as not to be overheard by the families sitting at the other windows. 'Perhaps with good cause.'

He had wanted them to visit Virginia – Maria Celeste, as he must now think of her – together, perhaps even pretend that they were still close, but Marina had disappeared after the ceremony.

'She thinks I've put my work ahead of you,' Galileo confessed, 'that you should be married.'

Maria Celeste wrinkled her nose. 'Married? I'm as happy as I can be. I have Jesus and you, why would I want any other man?'

'And your sister? Does Livia feel the same?'

Virginia took a moment before answering. 'She will learn to love what she has. We're safe here, that is most important.'

Galileo itched to tear down the partition and hug her. Despite everything she was saying, he felt as if he had lost his daughters for ever, and Marina as well. His eyes began to burn, and he dropped his head in shame.

'Father, we are all told to listen to our hearts, and that God will whisper our chosen names to us. I knew my name when I awoke one night for no obvious reason. The stars were shining, and newly carved inside me was the name Maria Celeste. I knew at once that it was an acknowledgement of your astronomy.'

Galileo looked up, still struggling with his emotions. He managed a small shake of his head.

'Yes. Would God have told me otherwise if you were a bad man?' Her eyes were imploring, deep and sincere. 'I think you have his blessing for your work.'

Galileo waited in a shadowed doorway outside the convent long enough to regain his composure and to see Marina slip into the meeting room. He turned to find Castelli, grateful for the monk's company because it forced him to return his attention to the job in hand. He could dwell on Maria Celeste's comforting words later when he was alone. For now, he welcomed a distraction. They made their way back up the hill, the twilight gathering around them.

'I swear this gets steeper,' said Castelli.

'Hurry up, I must reply to Apelles.'

'Don't you want to eat?'

'We can do that, too.'

'It would be wise to keep in mind who Apelles really is. Father Scheiner is well respected, Galileo. Do you really want to attack the Jesuits?'

'He leaves me no choice. Besides, he has made a serious mistake in choosing a pseudonym. Think: he has not published in his own name, nor through the Roman College. He is acting on his own, outside Jesuit rules.'

'You think this makes him vulnerable?'

'I do. Besides, if Grienberger ever asks, I'll simply say that I had no idea Apelles was one of his number. Your trouble, Castelli, is that you think too small; you have to look at the whole picture. The Jesuits are strong because they work together. But Apelles has given me the perfect opportunity to weaken their stranglehold on astronomy.'

'You've been forbidden to defend Copernicus.'

'But no one has said I can't attack the old ways of thinking.'

'You're a braver man than I.'

Galileo reached around Castelli's shoulders and drew him near.

'I've heard stories of tigers hunting on the Asian prairies. With infinite patience, one will stalk a herd. It watches and it waits for a foolish animal to stray from the safety of the others. Then it strikes.'

Galileo banged his fists together in front of Castelli's face, making him jump. Galileo laughed. 'Father Scheiner has separated himself from the herd. Now, I will pick him off.'

29
Leonberg, Swabia

The tiny cottage in Leonberg was more or less as Katharina had left it. The gate creaked a little more than when she had fled and someone had heaved a stone through the window, but apart from that it was much the same as before.

As Kepler unloaded their travelling cases from the carriage, a few onlookers gathered in their drab smocks. One of them cocked his head. 'Aren't you her son? Have you brought her back?'

Kepler ignored the question and hurried inside. The spiders had been busy, and there was a musty smell. On the floor, there were rat droppings. Katharina brushed them away while Kepler lit a fire from the few thin logs his mother had stored.

'Where do you keep your candles?'

'In the box on the mantelpiece.'

He opened the wooden container. There were a few stubs of wax. 'These are barely enough to read by for a single evening.'

'What do I need with reading? Are you hungry?' she said. 'I left some cheese somewhere, we can trim it up.'

'Mother, I'll go and buy us something from the inn. In the morning, we can get fresh eggs and bread.'

Katharina pulled a face. 'Have it your way. I can't get used to your city living.'

A violent banging on the front door roused Kepler from his sleep.

'Open up, Katharina Kepler.'

'That's him. That's Einhorn,' hissed Katharina as Kepler raced downstairs, his clothing dishevelled from an uncomfortable night on the floorboards.

The banging continued. She followed him, chewing her nails.

'Leave this to me,' said Kepler. He straightened his shirt, threw his jacket round his shoulders and ran a hand through his hair before opening the door.

Under different circumstances, Magistrate Luther Einhorn might have been considered a handsome man. His face culminated in a dimpled chin and his shoulder-length hair was thick and wavy. He held himself bolt upright, even though this meant he was too tall for the front door. Kepler stood his ground as the new arrival bent to look him in the eye.

'Stand aside,' said the Magistrate.

'It's not my mother you should be coming for but those who seek to slander her. I have written to Tübingen to explain this. We have nothing to fear from you.'

Einhorn's mouth twisted into a lopsided smile, destroying the façade of good looks. 'We've heard from Tübingen. That's why we're here.'

'To apologise, no doubt,' said Kepler but his bravado was ebbing.

'It seems that you don't have the influence you'd like to think you have.'

The Magistrate signalled to his men, who pushed past Kepler. The chains they carried rattled, and Einhorn raised his voice. 'Katharina Kepler, you are under arrest on the vehement suspicion of witchcraft. I have a charge sheet of forty-eight items that you must account for.'

The old woman backed away as the men advanced with the shackles held open. 'Surely it does not have to be like this,' she said. 'I have a silver cup, Magistrate, enough for you and your men . . .'

'Mother!' Kepler shouted.

Einhorn smiled broadly. 'I think the charge sheet has just grown to forty-nine. Take her away.'

Kepler had been in the cell for only a few minutes but already felt claustrophobic. In a place like this, even the strongest man's resolve would shrivel like an unpicked fruit in the midst of a drought.

At least there was daylight, spilling in from an open window criss-crossed by metal bars. There was a smattering of straw covering the stone floor, and a tiny plate of gruel that Katharina had not touched. She sat with her back against the wall, avoiding the shaft of sunlight. Her hands were manacled.

'What am I going to do?'

'Just tell the truth, Mother. They cannot harm you for that.'

The sound of the guards unlocking the door drew her to her feet. Einhorn ducked inside. He looked freshly coiffed and wore a heavy black cloak that buttoned at his neck and flared from his shoulders to his knees. He looked at Kepler. 'You must leave.'

'I demand to be present.' Kepler stepped close, lowering his voice. 'I know of your drinking session with her accusers.'

Einhorn's face betrayed nothing yet he said, 'Very well. In this instance you may stay. But say nothing.'

Kepler looked at Katharina's drawn face, hoping to transmit the minor triumph. 'Remember, Mother, nothing but the truth.'

'Yes, that's all we want Katharina, the truth.' Einhorn stepped forwards, eclipsing Kepler's view. The Magistrate's voice was conciliatory, almost friendly, and Kepler felt a stab of hatred.

A man edged into the room, carrying parchment and quills, and set about unfolding a small writing table and stool.

'What's he for?' asked Katharina.

'To record your confession.' Einhorn smiled. 'It's not just Ursula Reinbold who has levelled a complaint. The schoolmaster now says that you gave him the potion too, and that he became lame as a result.'

'He became lame because he was drunk and fell into a ditch.'

'Then there is the Haller girl.'

'She's just work-shy.'

'So you admit to harming her.'

'I never touched her.'

Einhorn circled Katharina, talking pointedly. 'That's not the story I heard. She claims that while she was carrying chalk to the limekilns you did reach out and touch her when no one was looking. It caused such pain in her arm that she had to be sent home.'

'Not even the brickmaker believed her story. She was just trying to avoid work.'

'It is also said that livestock become agitated when you're nearby.'

'Lies,' said Katharina.

Einhorn straightened himself, made a show of nonchalantly removing his gloves. 'You were brought up by your aunt, were you not?'

Kepler dropped his head, knowing where this was going.

His mother nodded, pursing her lips. She knew too.

'What happened to her?'

Katharina lifted her chin, said nothing.

'I'm waiting, Katharina.'

She clenched her fists.

'Katharina . . .' His eyes were burning. 'Tell the truth.'

'She was burned for witchcraft, as you well know,' said Katharina. 'I can't change what happened to her.'

'Indeed, you can't. You were raised by a witch.'

'I've done nothing wrong.'

Einhorn pulled the gloves through one hand. 'Do you know what prickers are used for?'

Katharina looked at her son with desperate eyes. He felt as though he would suffocate with frustration.

'They're hooked needles, about this long.' Einhorn held his two forefingers about eight inches apart. 'Designed to find the dead spots where witches have suckled their imps. If we were to use them on you, Katharina, how long before we'd find a dead spot?'

'You can't do that!' cried Kepler.

'Can't I? Tell me, Herr Kepler, you're a religious man. What would you say if I told you I have a young girl's testimony in which she says your mother told her that there is neither Heaven nor Hell, and that when we die everything is over, just as it is for the senseless beasts.' Einhorn crossed his arms, looking smug.

'Mother?' Kepler gasped, searched her face. She hurriedly looked away and glared at the Magistrate.

'I'm not afraid of you,' she said, though there was a quiver in her voice.

'Better to confess now and save yourself the terror of the torture chamber. Why prolong this agony?'

'I have nothing to confess.'

'Is it true what you said about Heaven, Mother?' Kepler stammered.

'Answer him, Katharina. Don't you want to go to God with a clear conscience?'

She said nothing.

'Very well. Perhaps you'll explain yourself tomorrow to the district torturer. Guard,' he called over his shoulder, 'I'm done here.'

Kepler rushed into the tavern the next day to buy food to take to the gaol. As he waited to be served, he noticed the scribe hunched over a plate of bread and cheese. The man became intent upon his meal when he caught sight of the astronomer.

Kepler went over. 'Something's wrong with all this.'

'I cannot talk about it. We must let justice take its course.'

'What justice is there in terrorising an old woman? For pity's sake, man, what do you know that I don't?'

The scribe looked away into the fire, conflict written on his face. Kepler pressed on. 'It's the easy way to make sense of the world, isn't it? Bundle up everything we don't understand and call it witchcraft. Sudden death, crops failing, animals dying: all robbing us of food. But it's a trap, because as soon as we start to believe in evil, we see it everywhere and become scared of it. You know my mother is not a witch; she's just a poor old woman. Why do this to her when she only has a few more years to live? Because she's ugly? Because she's been lucky enough to live a long life? Or because it takes the blame off everyone else for their failures? And when she's gone, what then? Babies will still die for no apparent reason. People will still go hungry. So, you'll need to find another witch and burn her too. How many do you have in mind? Two? Five? A hundred? Where is this all going to end?'

The scribe began to talk. 'There was not enough evidence to warrant your mother's death. Yet there was enough suspicion that the prosecution has been granted the right of *territio verbalis*. She will be shown the instruments. The torturer will explain what he will do to her with them. They'll hold the branding irons so close she'll feel the heat.'

'But he cannot actually torture her, can he? That's what you're keeping from us. The verdict is *territio verbalis*. All my mother has to do is continue to protest her innocence.'

'She'll confess; they all do. Then she'll be burned.'

Kepler sent the scribe's table crashing to one side. Food and beer flew through the air, sizzling in the fire. A pewter tankard clattered on the filthy flagstones. The scribe leaped up, tripping over his stool and into a nearby table, creating even more noise and mess.

'Hey!' shouted the barman.

But Kepler had vanished.

He bolted from the tavern in a headlong rush for the gaol. Dodging puddles, he nearly went sprawling as he collided with someone coming out of a doorway. He called out an apology and kept running. Heads turned to see him pass and geese scattered. He ignored them all, intent only on reaching his mother to deliver his message of hope.

But by the time he arrived, she had already been taken to the dungeon.

Only his mother's stubborn nature would save her now. If only he had seen through this sham earlier and warned her. He paced the streets, orbiting the stone tower of the gaol.

Next, he tried crouching down and bowing his head in prayer.

Dear Lord, please restore my mother's faith. She means no harm. She is aged and simple and sometimes confused.

But it seemed inadequate. He was shocked to even think it, but he doubted the slow turn of Heaven could save her today. He wanted to shout, cry, run away, break down the cell door. He wanted to do all these things and none of them. When he walked away to ease some of the tension that tore at him, he worried that he should be at the gaol. Yet, when he stayed put, he felt as though he would explode.

Einhorn swept towards him, his face thunderous. 'She's inside,' said the Magistrate without breaking his stride. 'What's left of her.'

Kepler rushed into the gloom, tripping on the narrow flight of steps that led to the torture chamber. The only illumination came from the red coals in the brazier and the glowing poker that was lodged within them. The room stank of sweat. The bare-chested torturer was hanging fearsome metal implements, covered in spikes, back on the wall. 'Get her out of here,' he snarled.

Katharina was bent double clutching her stomach. She looked so small.

'It's over?'

'Yes, it's over,' said the brute.

'Mother?'

She did not move until he drew close enough to support her. Her fingers gripped his wrist so lightly that he barely felt their touch. He wished for the strength to lift her in his arms and carry her back to the cottage, but he too was trembling.

Emerging into the daylight, they said nothing on the way home, just doggedly put one foot in front of the other and ignored the looks they drew. He listened to her ragged breathing and fought back his tears. She had beaten them. But this did not feel like victory.

When, at last, they reached home, he led her upstairs to her room. 'How did you resist them?' he asked as she curled up on the thin mattress.

Katharina began to weep quietly. 'I don't know.'

What warmth the day had possessed disappeared with the sunlight. Kepler climbed the creaking stairs to lay another blanket on his mother. She was awake as he entered the tiny room, staring at the ceiling.

'Are you going to ask that girl to marry you?'

It took him a moment to register the question. 'Mother, I hardly know her.'

She turned her canny eyes towards him, and he cursed himself for being so transparent. In the month they had spent in Linz preparing to return to Leonberg, he had thought his frequent visits to Baroness Starhemberg had been explained by his numerous excuses. First, he told his mother he wanted to gather a few more details, then to update the Baroness on the progress of her chart, then to give her a preliminary reading. His mother had seen right through him.

Perhaps he had carried home his elation the day Susanna kept him company while he had been waiting for the Baroness. Susanna asked him to tell her about the stars, and he launched into an explanation of the various zodiacal signs and their influences over the body. When he stumbled over his words, explaining how Leo ruled the heart, aware of just how fast his own was beating, she spoke up. 'Forgive me. I did not make myself clear. I refer to the true meaning of the stars, as you have discovered.' She grew a little embarrassed at his incredulous look, casting her eyes down and explaining that his eminence had been discussed at dinner the previous night.

'You want your children back, don't you?' Katharina was saying. 'You can't leave them in Kunstadt for ever.'

'Of course not.'

'What's her name? Will she make a good mother?'

Kepler smiled despite himself. 'Susanna. She's modest and thrifty, and she loves children.'

'What do her parents say?'

'She's an orphan. That's how she came to be Baroness Starhemberg's companion. They've raised her as if she were a daughter.'

'Not much of a dowry, then?'

'Mother! Such talk is unjustified. What may or may not transpire between Susanna and me is not to be reduced to a matter of money. Now, I don't want to discuss it further.'

'Alright. But you've got to talk to me – about anything – or my mind will start to fill with fear again.'

Kepler's eyes searched the room for inspiration. 'Though it's hard to believe at the moment, there must be harmony in the world; God's perfection cannot allow it to be otherwise. It must be a harmony so grand that it reduces all earthly woes to triviality.'

'I didn't want a sermon.'

'It's not a sermon. I'm trying to tell you something. Something I've not told anyone else.' He paused. 'I have searched for this harmony for years now and . . . and I've found it. Do you remember me saying that the Sun controls the movement of the planets? Well, for this to be true there must be a mathematical rule that links a planet's distance from the Sun to the time it takes to complete an orbit – that's called the period. I've found that the period and distance are a ratio – just as notes are in a musical scale. In the case of the planets, it's a little more complicated: the square of the period is always equal to the cube of the semi-major axis of the elliptical orbit. Do you understand?'

A faint smile tugged at Katharina's thin lips. 'I'll never understand the things you talk about, but don't stop talking.'

Kepler ran a finger over a patch of her blanket, describing an elliptical shape. 'This is an orbit, and the longest line I can draw inside is the major axis.' He ran his finger across the widest part of his pretend ellipse. 'Half of this line is the semi-major axis. Multiply it to itself three times and it equals the period of the orbit multiplied together twice. It is the wedge that can be used to drive open the mathematics of the solar force that governs the planetary motion.'

He paused for some sign of affirmation. When it did not come, he turned his head. In the Moon's pale facsimile of sunlight, he could just make out the rise and fall of her chest.

In the moments of calm that followed, Kepler understood his future. He could see himself putting all the things he had to do into action. With his mother safe, he must return to Linz and make a new life with Susanna Reuttinger, if she would have him. Kepler chased away the momentary doubt. Of course she would; he had seen the lingering way she looked at him.

For all her privilege, Susanna was lonely. She seemed to have accepted her orphaned status as her fault. Whenever he thought of her pale vulnerability, he wanted her even more. The shine in her eyes and the sparkle of her voice stirred him in a way that he had almost forgotten was possible. The more he made her laugh, the less it reminded him of Barbara and the more it sounded like a unique connection between him and Susanna.

He would retrieve his children from Kunstadt; perhaps have some more? And he would publish a book about the harmony of the universe, presenting his newly discovered law of planetary motion to the world. Then, at long last, he would finish *The Rudolphine Tables*.

30
Florence, Tuscany
1623

Galileo hamfistedly shuffled his papers again, banging them on the tabletop to force them into a pile. He was in the convent's meeting parlour. Raising himself with a grunt he peered through the iron bars of the nearest window, scanning the empty room beyond.

Where is Maria Celeste? And when did I stop thinking of her as Virginia, the name I chose for her?

He had so much to tell her.

Six months ago, the world had changed in a puff of white smoke from a Vatican chimney. Ignorance and doubt and fear had all been swept away as Cardinal Maffeo Barberini became Pope Urban VIII.

At word of the accession, Galileo had ransacked the trunk of letters in his hallway, finding the correct bundle tied in a silk ribbon, and almost torn the fragile papers in his haste to reread them. They were the letters that the new Pope had written to him years before, when still a cardinal, praising his astronomy and his intellect. They were just as Galileo remembered. Urban referred to him as a pious man of great virtue whose astronomy improved the lives of others.

Galileo had wasted no time in sending his congratulations to Urban. He had considered a direct letter based on these past exchanges, but then he thought the new Pope was more likely to respond to a show of excessive humility. As divine providence would have it, Galileo had recently steered Urban's nephew through a doctoral degree so he used the young man as his messenger.

A few weeks later, a courier presented a missive that made Galileo punch the air in triumph. Keen to continue and even strengthen their correspondence, the Pope had invited him to the Vatican. Galileo laughed at the thought of the consternation this must have produced in the Inquisition offices, not to mention the Roman College. Everything could change now that the Pope, supreme ruler of all Catholics, was a supporter of Galileo.

Galileo had arrived back from Rome yesterday evening, and he had a story to tell.

Come on, Maria Celeste!

He dropped back into the uncomfortable wooden chair and impatiently studied his side of the room. It was clean, but unrepaired holes in the walls revealed the hair and straw of the inner binding. Even the light seemed old in here.

There was movement behind the metal grille. Maria Celeste was by the far door, carrying a small bundle. He pressed himself closer to the wall and held his palm to the metal railing. Their fingers locked. 'You're cold,' he said.

'Sister Arcangela wanted me to dowse her again.'

Arcangela – Livia's chosen name. 'How is your sister?'

'She's still in her bed with the fever I wrote to you about.'

'Is it that serious?'

Maria Celeste spoke with deliberate care. 'There are those who might have shrugged off the illness sooner.'

'I see.'

'Here are your collars. I have mended and bleached them.' She passed the garments between the metal bars. Galileo stacked them with his papers.

'Now let us talk of you and your adventures. It is unforgivable of me to keep you waiting when you have so much to tell. I've missed you this past month, Papa. Tell me, did you meet him? Did you meet the new Pope?'

'I did. We walked together in the gardens of the Vatican.'

'Forgive me for being curious, but what is he like?'

'Everything you could wish for. He retains his great interest in my astronomy. Things will be different this time. Every day for five days we met to discuss philosophy. Just him and me.'

For their first meeting Urban VIII had approached on horseback, the colour of his cassock matching the white hair of his stallion. Galileo had felt his heart accelerate at the majestic sight. Had he not been wearing an expensive new tunic, he might have felt intimidated.

The Swiss Guards kept a respectful distance but were watchful as Urban drew his steed to a halt, patted its neck and slid from the saddle. The Pope looked younger than his fifty-five years. Galileo fell to his knees and kissed Urban's outstretched hand.

'Did you hear? Bellarmine passed away,' said Urban.

'I did, Your Holiness.' Galileo stood up.

'Died penniless. Managed to give everything he had to the poor,' said the Pope. 'I expect I'll have to commission a statue of him.'

Galileo said nothing.

Urban had a square goatee that protruded some two inches from the bottom of his chin. His moustache was waxed into horizontal points. He moved steadily, each step seemingly designed to savour his new position in the world. Galileo fell in beside him.

'You haven't forgiven him for the anti-Copernican edict have you?' asked Urban.

Galileo watched a pair of birds wheel through the sky, choosing his words carefully. 'I have a lot to thank Cardinal Bellarmine for. Without him, the edict may have been harsher. As it was, Copernicus wasn't banned, only corrected.'

'Rome is different now.'

Galileo turned, unsure he had heard correctly.

Urban's round eyes were burning. 'I will turn Rome into a hub of human education, both spiritual and philosophical. I will send missions into Europe and far beyond to take our teachings to the world. But to do it, I need help.'

'I am your humble servant as you know, but, Your Holiness, you have the Roman College on your doorstep.'

'Oh, quite so. I've studied with the Jesuits and value them above everybody else, yet I also understand the drawbacks of their system, their resistance to change.' Urban was watching Galileo closely.

'The Jesuits are the arbiters of Roman knowledge, Your Holiness.'

'Oh, come, don't fence with me. We both know that you skewered Father Scheiner over his interpretation of the sunspots.'

Galileo's heart accelerated again. The Pope was flattering him: he wanted something. 'Your Holiness, the good philosopher flies alone like an eagle, not in a flock of noisy starlings. And though the lone voice may struggle to be heard, yet may he reach heights that no starling can ever imagine.'

'And I am struck by one of your thoughts in particular: that you champion experimental evidence over the wisdom of the ancients.'

Galileo stopped dead. 'You have read my latest work about the comets of 1618?'

'*Il Saggiatore*. Yes, I have it read to me at mealtimes. I'm not sure there is anyone else in all Europe who can present a polemic like you. I am captivated by your new way of learning.'

'We should believe only what we can verify by experimentation. Nothing else is credible. Deduction and logic are second rate compared to the experience of our eyes and the wit of our mathematics. We see nature around us; let us truly investigate it. In this way, we can bring ourselves closer to God.'

'It seems so obvious in hindsight.'

'If I may say so, Your Holiness, the greatest ideas always do.'

Urban walked on, forcing Galileo to match his pace. 'There is one thing that surprises me, Galileo. The Roman College now say that they have measured all three comets and that they are celestial objects, able to move between the planets. Yet you maintain in your book that they are atmospheric phenomena.'

'Your Holiness, one only has to spit on the floor to see sunlight thus reflected. But only a fool would think he has discovered a new star.'

Urban inclined his head in a sideways nod. 'There is one thing that *Il Saggiatore* does not mention: the system of Copernicus.'

Galileo could not help but glance over his shoulder. The Swiss Guards were a dozen paces behind. Nevertheless, he lowered his voice. 'The edict of 1616 . . .'

'If I had been Pope then, there would never have been an edict.'

Galileo found it hard to convince himself that he had heard correctly.

'Oh, the look on your face, Galileo . . . I find nothing wrong with Copernicus's ideas – so long as they are confined to hypothesis, never spoken as truth. There are still too many in Rome who believe the planets are moved by angels to talk of it as anything else. Claim that it is just a mathematical trick to achieve the correct answer rather than assert that this is the true arrangement of planets and all seems reasonable to me.'

'What of the Jesuits and their favour for the Tychonic arrangement?'

'I think you are fencing with me again.'

'Forgive me. As we both know, Tycho's system was never more than an ugly compromise.'

'A stepping stone to Copernicus. If I am going to send my priests and monks off into the world, I need to arm them with the best weapons against ignorance that I can find.'

Galileo felt a shiver pass through him. 'A book,' he whispered. 'You need a book that anyone can read arguing for Copernican astronomy.'

'Know anyone who might like to write it?'

'Your Holiness, it would be my honour to begin at once.'

Urban smiled. 'Perhaps I will commission a statue of you one day, Galileo.'

Galileo nearly laughed at his daughter, her eyes were as wide and white as the full moon and her mouth was parted in astonishment. He told her, 'I'm going to write it in Italian, not Latin, and in ordinary language for the common man to read. They will be my judges. I'll include no mathematics in the actual book but all my arguments will be based on my lifetime of observations and calculations. I'm going to frame the book as a discussion between three philosophers, taking place over a number of days. One philosopher will be a Copernican; the other will be a foolish Aristotelian; and the third will be an undecided but reasonable man. On the first day, I'll draw the battle lines: the Aristotelian will describe the Earth as being fundamentally different from all the other celestial objects; and the Copernican will point out that the telescope shows this is not true – that the Moon is Earth-like and probably so are the other planets. On the second day, I'll talk about the twofold motion of the Earth; its daily rotation to give us night and day, and its annual orbit to give us our year. Day Three will see the introduction of the sunspots, while on the fourth and final day they will discuss the tides. By the end of this day, our agnostic will have no choice but to side with the Copernican. There will be no doubt left. At last, I have been given the chance to prove myself.'

31
Linz, Upper Austria
1627

Kepler was dreading the journey to come: several days in a freezing carriage skidding across muddy roads that had hardened into glass. He had tugged on two pairs of hose that morning in preparation, yet still felt cold at the thought.

Susanna secured the last few buttons on his jerkin and smoothed the material along his torso. 'After all these years, I thought you'd have fattened up.'

'And, after all these years, my wife, I thought you'd have stopped worrying.' He drew her close and kissed her, marvelling at the thrill in it. A thrill that fourteen years of marriage, six children – three surviving – had yet to erode.

They had celebrated the marriage with a feast at the Sign of the Lion Inn where Emperor Matthias's representatives had caused a stunned silence when they presented Kepler with an extravagant goblet as a wedding present.

After all these years, he could no longer bring to mind the dress she had worn, except to say that it had been beautiful, but he vividly remembered looking into her hazel-flecked eyes and thinking to himself, *I am reborn.*

'I'm sure everything will be just fine,' she said, relaxing into his arms.

'The tables are twenty-five years late,' Kepler said wearily. 'There have been two emperors since Rudolph.'

'So why must they be called *The Rudolphine Tables*? Why not *The Keplerian Tables*? After all the work you've put into them.'

'It is those in power who are remembered, not the people who do the work. I wonder how Emperor Ferdinand will receive me.'

'Your exile from Graz was almost thirty years ago; he'll have forgotten all about it. You've said it yourself: he was little more than a boy. And no man could have worked harder than you. With all your other books you've not been idle. I'm more worried about the journey. There are soldiers everywhere.'

'The worst of the fighting is far from here, and I have all the necessary travel papers. I'll be home with gingerbread before Christmas.' He released her and, peering in the hall mirror, fixed his large floppy beret. The winter light gave the polished metal plate little illumination to work with. He appeared but a shadow, with a flowing beard of grey.

Susanna held open a fur-lined coat and slid the weight of it around his shoulders. He pulled on gloves, covering the liver spots that now mingled with the pox scars on his hands.

'The children will miss you.' She pulled him tight again. 'I'll miss you.'

For a moment, Kepler feared he was doing the wrong thing. He suddenly doubted his own promise to return by Christmas. What was he doing marching into the Palace of the most powerful Roman Catholic after the Pope? Ferdinand knew only too well Kepler's staunch views against papal authority. He had not appointed Kepler to be his imperial mathematician but inherited him out of necessity because the astronomer was still compiling the long-promised tables. It was a task that Kepler had pursued hesitantly, allowing himself to be freely distracted by his own quests.

If he did not set off now, he never would. Kepler kissed Susanna again as he opened the door.

'Johannes, wait.' Susanna handed him his cushion.

'What would I do without you?'

At the city gate, the carriage was flagged to a halt. A scruffy guard smelling of beer cast a suspicious eye around the compartment. Kepler proffered his papers. Although he doubted the man could read, he trusted the red wax of the ducal seal would suffice.

The crack of fracturing wood split the air. Kepler looked up to see that another city guard had smashed open a newly arrived couple's travelling chest and was rooting through their clothes. In grave triumph, he lifted a Lutheran Bible from the muddle. Swaggering to the brazier, he dropped it into the flames.

Kepler's guard dropped the papers back inside the carriage and turned to help his colleague. Together they beat the couple, making no distinction between the man and the woman, then pushed them on their way, bleeding and with their clothes bundled in their arms.

At night, curled into one strange bed after another as the journey continued, Kepler fought back the memory of their broken faces. And he dreamed of Ferdinand, as he had known him back in Graz; just twenty-two but puffy-faced with a drooping nose. Ferdinand used to twirl his waxed moustache in a comic affectation of boredom as he watched the Lutherans being converted one by one, but it had all changed when Kepler stepped before the panel. Now Ferdinand was glowering at him, anticipating the disobedience of his mathematician – who did not disappoint. Kepler always awoke at the moment he voiced his rejection of Rome, clammy and panting.

Susanna was right, he kept telling himself, it had all been a long time ago. But in the back of his mind, he knew that if he could still remember that day, so could Ferdinand.

Solid banks of white clouds covered Prague, as if God had blocked the city from His view. Armoured men roamed the streets, stopping people at random to check their destination and intentions. Kepler's carriage took its place in the queue of traffic on the bridge where imperial guards were searching all vehicles.

Today there were some grisly additions spiked and mounted on the balustrade. Twelve weathered skulls glittered in the frosty morning – twelve of the most infamous 'traitors' executed by

Ferdinand upon his ascendancy. On that day, twenty-seven Protestant leaders had lost their lives: most beheaded; three hanged. The heads of the most hated offenders were displayed on the bridge as a grim warning to all who held Lutheran beliefs.

Kepler had been reluctant to believe the news when he'd heard it in Linz. Now here was the ghastly confirmation. He averted his gaze whenever the halting progress of the coach brought another desiccated skull hovering alongside the window. However, one proved impossible to ignore. A rusty nail protruded from the cracked forehead, supporting a few withered fibres that had once been a tongue.

Kepler's stomach fell away. The gossip in Linz had specifically described this mutilation. That tongue had negotiated his introduction to Tycho Brahe; it had appeased the Danish astronomer when a fever had rendered Kepler irrational; and finally it had brokered the reconciliation when Brahe had moved to Prague. Had it not been for that tongue, Kepler would not be here today presenting *The Rudolphine Tables*. The tongue and the skull belonged to Jan Jessenius.

Jessenius had inspired an exquisite hatred in Ferdinand because he had united the Bohemian estates into a coherent Lutheran force. After capture and execution, in a disturbing parody of his profession, the anatomist's body had been quartered and hung around the city.

Could it have escaped Ferdinand's spies that during Kepler's time in Prague he had been a close associate of Jessenius?

The carriage passed through the search without incident and rumbled on through the traffic to the market square. They passed the town clock, the mechanical Apostles silent for the time being, and approached the overwhelming spires of the Church of Our Lady Before Týn.

Kepler had the driver pull up alongside another waiting carriage. He climbed down to the pavement and straightened his cloak, then reached back inside to retrieve a large leather-bound book.

'I won't be long,' he said to the driver.

Inside, the church was quiet. A woman sat in the back row gazing at the magnificent altar. Kepler followed her gaze and his jaw dropped.

The portrait of Utraquist leader Georg von Podiebrad that used to hang above the altar had gone, replaced by a statue of the Madonna. The Virgin Mary's tall figure was adorned with a crown and surrounded by an aureole of gold. In her arms, Jesus was blessing his mother.

Down on one knee at the base of the altar was a broad-shouldered man, dressed head to toe in black. His head was bowed in prayer. Kepler bobbed his own head at the altar and made his way into the transept, to a particular marble slab. It lay horizontally on the floor, flush with the other flagstones, and was inscribed with letters that had been painted in gold: TYCHO BRAHE.

Kepler kneeled before his old master and lifted the book. 'It is done,' he whispered. 'Your observations, my mathematics – just as you always envisaged. Your life's work will live on now and provide the most accurate planetary tables in all history. You helped me unlock the elliptical orbits and the movements of the planets. You may not have appreciated it at the time but perhaps now, looking down from above, you can see them too. While our own world becomes more dissonant every day, thanks to you I have seen the greater harmony of the cosmos.'

Kepler laid the book on the gravestone, clasped his hands and prayed silently.

When he returned to the nave, the book back under his arm, the man at the altar was just getting to his feet. Something about him aroused Kepler's curiosity and he hung back, watching from the shadows. The man was even more powerful than at first sight, martial in his bearing. Kepler guessed he was in his mid forties.

The man crossed himself and strode down the aisle. As he reached the rear pews, the woman dropped to her knees before him and kissed his hand. He accepted the gesture and moved on. As he was about to leave, a bent old man shuffled through the doors and instantly came to attention. The man nodded his appreciation but did not break his stride.

'Who was that?' Kepler asked the driver once outside, indicating the carriage that was leaving.

'That, my friend, is General Wallenstein, saviour of Prague, defender of Catholic pride.' The driver melodramatically waved his arm.

Conqueror of vast tracts of Protestant lands, thought Kepler. Wallenstein was a war hero here because of his military successes against the Protestant estates. By way of reward for his victories, Ferdinand had given him the Duchy of Friedland, north of Prague.

Kepler watched the carriage trundle away. As it passed, pedestrians would occasionally look up and point, or applaud.

Kepler heaved himself into his own vehicle. 'To the palace, please.'

Turning up Hradčany Hill the palace faced them, looking more like a fortress than a palace. With guards stationed at the locked gates and on the ramparts, there were even lookouts on the astronomer's tower that rose from Rudolph's former Arts Chamber.

The soldiers were not needed for defence – the city was safe now – so this was a show of strength, a display that Prague was the iron heart of Ferdinand's Empire, united in its Catholicism.

Kepler imagined a noose slipping around his neck.

Children appeared to be running the Imperial Court now. Worse still, children pretending to be adults. The austere cut of their clothing seemed at odds with their youthful faces and they looked more concerned with their appearance than with government.

Glancing upwards in the reception hall, Kepler expected to see those magnificent arches, but the light was so bad today that the exquisite tracery was lost in the highest vaults. Or was it his eyes?

'Please wait here. Do you have the book?' The court official was coldly formal. Although he put Kepler on edge, it was preferable to insincerity. He handed over the *Tables* and resigned himself to a long wait.

Watching the toing and froing, Kepler despaired. The officials grimaced at each other and postured in a manner that Kepler had noticed was endemic in young men before the comfort of accomplishment softened their intensity. It was no wonder that war was upon them. Not a single one of them here was old enough to foresee the consequences of their actions. Every decision would be taken with self-righteous declarations of piety, designed to absolve them from the consideration of any suffering it might cause to others. Thus, the counter-reformation continued its bloody march.

Ironically it had been Kepler's excommunication that saved him when orders banning Lutheran services had arrived in Linz. Pastor Hitzler had been imprisoned; fines were raised on burials and marriage services for Lutherans; and, of course, the schools and churches were closed. Emigration was offered as the only option for those unwilling to convert: but to where? Catholicism had the momentum again. Only the Protestant heartlands were safe these days, or England.

Kepler had slipped through the net. He had made money by soliciting gifts from European nobles, sending them lavishly bound copies of his various books and, occasionally, this very Court remembered to pay him.

His mind drifted to Susanna and his cherished family; his second and surely his last: Cordula, with her six-year-old girlish chatter; four-year-old Fridmar and his thoughtful – if hesitant – sentences; and finally the somnolent Hildebert, who at three should have been making a lot more trouble than he was. Kepler felt a pang of homesickness and wondered what he was missing. He was becoming increasingly aware of time's passing.

At fifty-six, he did not feel particularly old, but he certainly did not feel strong.

'This way, please.' The court official interrupted his thoughts.

From inside a nearby anteroom another official opened yet another door and wafted through, announcing Johannes Kepler.

Ferdinand was jammed into a wooden throne. The tiny eyes and long nose were those of the boy Kepler remembered, but the once sandy mop of hair had dulled and thinned, and the fading ginger goatee unnaturally lengthened his face. His fleshy body was hidden beneath a gigantic robe of black and gold making his head look too small.

A pair of fawn-coloured pugs engaged in a friendly skirmish near the corner of the room. No one else paid them any attention, but Kepler found them distracting. He dragged his attention back, bowing as deeply as he could. 'Your Majesty, you are most gracious to receive me.'

'It is an honour to have you here at last.' Ferdinand's mouth twitched into a smile. 'It is my understanding that you always enjoyed living in Prague.'

'I did, Your Majesty.' He resisted adding that in those days it had been a place of stimulating diversity.

There were a number of other people in the room. A scribe sat at a little desk, quill in hand, and advisors hovered near the Emperor; one of them held *The Rudolphine Tables*. Another man was dressed head to toe in a black Jesuit cassock.

Ferdinand followed his line of vision. He said, 'Father Paul Guldin, I believe you two know of each other.'

Guldin? Yes, that name did sound familiar.

'I helped furnish you with a Galilean telescope, Herr Kepler.'

'Of course, my apologies for my lack of recall. Without that telescope, I would have fallen behind.'

'It was my pleasure. A man of your learning must have the instruments to make the most of his gifts.' There was genuine warmth in Guldin's eyes. If anything, it made Kepler more nervous.

The aide handed the book to Ferdinand, who turned the wide pages reverently.

'Father Guldin informs me that this is the greatest work of positional astronomy since *The Alphonsine Tables* from the thirteenth century. It is a work that could be used for centuries to come.'

'They are the first to correct for the deflection of starlight by the thicker layers of atmosphere near the horizon, it is true. But let me say that the undoubted quality of this book is due to the observations of the late Tycho Brahe. His observatory was unsurpassed.'

'You are too modest for a man of your abilities. Your years of painstaking calculation match those of Brahe's observations.'

'I am a humble computer, nothing more.'

'Nonsense. You are the greatest living astronomer. Not only have you charted the stars, but you have supplied prognostications of remarkable accuracy.' Ferdinand shifted his great bulk. 'Would it surprise you to know that I too can play the prophet?'

Kepler was nonplussed by the turn of the conversation. Ferdinand smiled and continued. 'Let me tell you what I foresee for you. Firstly, in honour of your great work that glorifies my departed cousin Rudolph, I foresee that you will receive a total of four thousand gulden from the cities of Nürnberg and Ulm.' He glanced at the scribe, who noted down the order. 'Secondly, you will take your place in my court; you have been in the provinces for too long. Thirdly, I foresee a comfortable teaching post, so that you can spread your wisdom to the generation of astronomers that will succeed you.'

Kepler stared at Ferdinand. 'I am to stay in your service and remain in imperial lands, Your Majesty?'

'You are a trusted and loyal servant, closer to us now than you have ever been. It is time to bring you fully into our heart.'

The dogs caught Kepler's attention again. Their play had become more boisterous and they were growling at each other. A heavy goblet thumped into them, eliciting yelps. They skulked off in opposite directions, heads down.

'Get them out of here,' shouted Ferdinand in a fit of pique, lowering his throwing arm. One of the aides hurriedly shooed the dogs out of the room.

Kepler hastily expressed his thanks and backed away. He expected to be accompanied by the anonymous official who had shown him in, but it was Guldin who moved to escort him.

'Let us discuss your move to Prague,' he said, when they reached the corridor.

The reality of the Emperor's offer was slowly beginning to dawn on Kepler. This was the vindication he had yearned for in Graz, what he had hoped for in vain on the day of his expulsion. Of course, back then, Ferdinand had been too young to understand.

Guldin led him to a small room in which a table was laid with refreshments. Guldin poured two drinks and slid one across.

'To Prague,' said Guldin.

'To Emperor Ferdinand.' Kepler drank deeply.

'There are a few details that the Emperor has asked me to go through with you, so we can finalise this agreement.'

'Of course.'

Guldin said, 'We are aware of the rather strained state of affairs between you and the Lutheran Church.'

'I believe we are past strained. The Lutheran Church and I are broken. But it is of no concern to me; I am at peace with God in my heart.'

The ghost of a smile crossed Guldin's face. 'It is of concern to His Majesty that you appear to be outside religion.'

The golden aura that had been growing around Kepler since the audience turned black and crumbled. 'This is another attempt to convert me, isn't it? Isn't it?'

'You cannot work for the Emperor unless you turn to Rome. Soon after you were excommunicated, you referred to the Lutheran Church as a gelding animal basking among roses and staring at its enemy. An enemy, you said, who would soon cause its death. Those were your words, were they not?'

'I never made it clear who the gelding was. It could just as easily be the Roman Church, or the Jesuits.'

'Come, Johannes, that has never been your style. I've watched you long enough to know that the only anger you have is directed inwards. I understand that feeling.' Guldin interlaced his hands and leaned on his elbows. 'I was once like you: undecided.'

'I am not undecided.'

'Johannes, I am your friend. Friends do not have secrets. I'll trade you one of mine. I was baptised a Protestant. Yes, it's true. But I began to have my doubts: all that self-congratulation and superiority. Even then, to turn to Rome was the hardest decision I have ever taken but now look at me. The Jesuits have given me the strength I never had before. You could join us, and we will carry your ideas across the world. We know that your interpretation of Copernicus is correct. We trust your elliptical orbits as the true astronomy.'

'But the papal edict . . .'

'The edict was forced upon us by that blundering fool, Galileo. His pathetic ambitions have hobbled us for years. But we have been patient, and even as we speak there are moves to silence him once and for all. He has opposed the Jesuits once too often. You only need to read his works to realise that Galileo doesn't understand Copernicus as you do, and as we do. He still believes in circular orbits, even though you and I know they cannot work. He still believes comets are atmospheric phenomena. Your books are the future. Join us and we will help you to secure your place in history.'

Kepler looked squarely at Guldin. 'I will leave this life as I entered it: a member of the catholic church.'

The Jesuit's eye's widened.

'Do not take that to mean the Roman Catholic Church. I speak with the original meaning in mind. The one derived from the Greek *katholikos*, meaning universal. My catholic faith embraces all of Christ's followers with equal respect. Yes, respect – the one thing each Church denies the followers of the other. Respect between fellow Christians is all I ask.'

'But there are such great differences of opinion between the Churches.'

'What we agree on is far greater than what we disagree about. Divisions betray the weakness of man, not the will of God. Yet the fact that we do disagree is all that seems to matter. There will always be quarrels in any family. There are differences of opinion between the burghers of our great cities, yet they all regard themselves as citizens. We were all baptised as children of God. Some of us follow His Holiness; others the teachings of Luther; and yet others follow their own conscience.'

'Johannes, you're going to throw away your future. You will be destitute if you walk away from this opportunity.'

'Yes, I will, but, even as I beg in the streets, I will still be true to who I am. My work is finished here.' He fastened his cloak and rose. 'It's time for me to go and buy gingerbread.'

Kepler strode through the Palace with new purpose. Although he doubted that he would find a carriage bound for Linz at this time of day, he might get as far as Tabor before nightfall. That, at least, would mean he did not have to spend any more time in Prague. Finding his carriage in the yard, he asked his driver to take him back to his lodgings. He would collect his things, buy the children their gift and go at once to the coaching inn in the south of the city.

The carriage rocked its way down the hill. It carried him across the bridge and through the confounded checkpoint, then it turned to track the misty Vltava up past the university to what had once been the Jewish quarter. Unexpectedly the carriage drew to a halt.

'Why have we stopped?' demanded Kepler, sticking his head out of the window.

'To test a suspicion.'

'What are you talking about?'

'Look behind you. They've been following us since we left the Palace.'

Some way back, another carriage had also drawn to a halt. The horse shook its head; its driver sat still as a statue.

What now? Spies? Assassins?

An incandescent rage burst inside him. He flung open the door and marched towards the other vehicle, pulling in great lungfuls of the bitter air and sending clouds of vapour billowing from his nostrils. He was within ten feet of the carriage and preparing to launch his tirade, when the carriage door opened and a well-dressed man emerged. He doffed an extravagant, wide-brimmed hat.

'Apologies for the somewhat unorthodox approach, Herr Kepler.' He was slender and youthful in appearance, with thin lips, although not unkind, and dark eyes; the kind Kepler found easy to believe.

The man wore a satchel, into which he reached. 'I have something for you, something you may recognise.' He proferred Kepler a battered sheet of paper. Yellowed with age and frayed at the edges, but unmistakable: it held the diamond-shaped chart of a horoscope that Kepler had done years earlier. He looked up. There was a hint of amusement on the man's face.

'Where did you get this?' demanded Kepler.

'My master has owned it since you wrote it for him.'

Kepler pieced together his memories of the event. There had been an anonymous customer who had requested a horoscope. He had sent a messenger to find Kepler and paid handsomely for the job. As Kepler had charted the mystery man's nativity, he had found it to be striking in its leadership qualities, similar to that of the former English queen, Elizabeth.

'That must have been more than fifteen years ago. Your master, does he still live?' asked Kepler, looking into the man's dark eyes.

'He does.'

'Was it you who approached me last time?'

'No, sir.'

There was writing on the horoscope after Kepler's original notes. Someone had recorded various personal events and linked them to Kepler's prognostications, forming a patchwork biography. Kepler read the spidery hand, but the notes were too cryptic to make much sense. 'What does your master want?'

'He would like you to update the chart.'

A moment ago, Kepler had been set on speeding back to his family and leaving this former life behind once and for all. Now, he was wavering. He hated the part of himself that was so easily ensnared by intrigue but, if he spent just one more night here, perhaps he could resolve this loose end, as well.

'Who is the person?'

'I'm not at liberty to tell you. The arrangement is the same as before.'

The money would certainly come in handy. Kepler deliberated a moment or two longer, then said, 'I'll do the work tonight.'

'Bring it to the Half Moon Tavern tomorrow at noon.'

32

What had started as a light drizzle had become a downpour, and Kepler was soaked by the time the black timber frame of the Half Moon Tavern came into view. He opened the door, and the reek of damp clothes, hair and beer hit him at once. He tunnelled his way into the press of people and began to search for his man. There was a tap on his shoulder.

'Do you have it?' Those dark eyes flashed.

Kepler handed over the scroll.

'Wait here.' The man disappeared.

Kepler was just about to give in to the barman's stare and order a drink when the slim figure reappeared.

'Come this way, please.'

Kepler followed him through a doorway, then down a gloomy corridor to a back room. Inside sat several men, all wearing swords and fine clothes. Their conversation stopped. Kepler saw his horoscope spread across the table. Facing him was the broad-shouldered man from the church. He was smiling. It looked almost incongruous on such a military face. He stood up. 'I am . . .'

'General Wallenstein.' Kepler completed for him.

The General stepped around the table. 'Are you surprised to see me?'

'Your chart makes perfect sense now.'

'I was an ambitious young man when you drew my first horoscope. Now, I have achieved everything you said I would. Naturally I want to know: what's next?'

'My work is for people who understand philosophy, not for those who believe that I can pluck future eventualities from the heavens.'

Wallenstein recited from memory: *'If he pays proper attention to the events in the world, he will attain the high honours he seeks, along with great wealth.'* He continued: 'You wrote that in your original horoscope, and I took it to heart. I know the value of your advice and the part the individual plays in making his own fortune. You are a shrewd observer: one eye on the stars; the other on those around you.'

'I've seen farmers grow pumpkins in Linz. They tie ribbons around them to mould them into fanciful shapes, but no one would say the ribbons make the pumpkins grow. So it is with the stars – they may influence us but they do not predetermine us,' said Kepler.

'I'm not sure I like being thought of as a pumpkin but I take your point.' Wallenstein crossed his arms. 'I heard about your meeting yesterday. It confirms what I have long thought and admired about you. You're an individual, intent on following your own path, even if it leads you into difficulty. That applies to me too. Oh, I see in your face that you're sceptical of our similarity but consider this: I want power and you want knowledge. We're both willing to do whatever we must to achieve our goals on our own terms. So how different are we, really?'

Kepler paused. There was expectation in the air, though for what he could not yet tell. 'Is the new horoscope to your liking?'

'Indeed, it is, but I want more. Let me ask you this: do you really believe in your new astronomy? Or is it just a mathematical trick?'

'I believe most truthfully that God has revealed the working of his heavens to me. When I contemplate the stars, I find God – something I cannot say to be true when I look around Earth. Years ago, a new star appeared near Jupiter and Saturn. I witnessed it from the roof of the Imperial Palace and wondered what new age it foretold. I now believe that when this war is over, we will enter an age of reason, an age when mankind will step beyond superstition and investigate the natural world with rationality. We will believe nothing unless it can be measured or observed.'

'Then shall we throw away our beliefs in the horoscopes altogether?'

'Perhaps, but we must be careful. There are effects coming from the stars that we can measure. The tide ebbs and flows in response to the Moon. The planets circulate in response to the Sun. If the mighty oceans are moved by forces from space, so too – surely – must the fragile human soul. Perhaps we just do not yet have the means with which to measure such subtleties.'

'How do you spread this wisdom?'

'I've written many books to explain my insights and the technical details of how to derive my planetary laws. Now I must bring it all together in one volume. The man who originally taught me Copernicus has turned his back on it. Mästlin's book, *Epitome Astronomiae*, is the standard text in all universities yet it expounds the old ways of thinking. And he's not updated it in decades to include my work. So, I will write my own *Epitome*, that of Copernican astronomy, and distribute it. Once people see that and start using *The Rudolphine Tables*, they will see the accuracy of it, and appreciate its simplicity, its elegance.'

'And how will you do that when you're being chased from one city to another by Ferdinand's Protestant witch hunt?'

'I'll do what I can. You yourself are an architect of that witch hunt.'

The General shook his head. 'I'm a military man. I fight armies, not individuals.' He pursed his lips. 'I could provide you with a printing press. No more struggling to find a publisher. Would that interest you?'

'Of course, but I'm wise enough to know there will be a price attached.'

Wallenstein laughed. 'I said you were shrewd. The price is that you become one of my advisors. Money and power interest me. Art and science interest you. We are perfectly matched. Each needs the other yet neither will admit it. I can supply you with a place to live in peace, under my protection. In exchange, you can supply me with some of your shrewd judgement.'

'Are you offering me patronage?'

'Patronage and protection. No one has ever known what to do with you: not Rudolph; not Matthias; and certainly not Ferdinand. They've all tried to control you because of your religious beliefs. But I know exactly what to do: protect you and then leave you alone. Your beliefs are of no concern to me. You need no stick and no carrot to work. You are the prophet of the new age. I'll give you my offer in writing. It will mean that you move to Sagan, in Silesia. You'll have to adapt to a new country and a new language, but you'll be free to work on whatever you wish. Think on it. There's a place for you and your family there, if you'd like it.'

The messenger rummaged in his satchel and pulled out a sheet of parchment, covered in writing. Kepler accepted the document, read the words. They reiterated everything Wallenstein had said. This was not how Kepler had imagined this day unfolding at all.

He looked from face to face, hoping to see some flicker that would guide his decision. Perhaps the guarded smirk of a trap about to be sprung, or the eager twitch of duplicity.

But Wallenstein and his men met his gaze, their faces unreadable. Either they were master bluffers, or possibly – just possibly – they were telling the truth. Either way, this was Kepler's decision alone.

He thought of Susanna and the children having to pack up and move when all they had known was Linz. He was the nomad, not them. They were comfortable in the house even though the city was becoming increasingly dangerous. Could he start again at fifty-six, or rather fifty-seven, as he would be two days after Christmas? It was a daunting prospect, but as he identified that shadowy feeling, so a spark kindled inside him to chase it away. It was the same glow of excitement he used to feel when contemplating the future. In his mind's eye he could see his completed textbook, feel it in his hands. He should really go home and discuss this all with Susanna. Instead, he looked at Wallenstein.

'I accept.'

33

Rome, Papal States

1632

Pippe knew something was going to happen. He could feel it in the same way he sensed a coming thunderstorm during August: by the pricking of the hairs on the back of his neck. What he could not sense was whether it would help his own agenda in today's meeting.

The seventy cardinals sat in a horseshoe formation in the large meeting hall, along with the various Vatican ambassadors visiting Rome. Their papers were placed on the tables in front of them. Their voices echoed from the bare marble. The canny ones heeded those reflections as warnings. Whenever they could hear themselves rebounding from the walls, it was time to calm down.

Not so the Vatican ambassador to Spain.

The voice of Ambassador Borgia issued from his pudding face to bounce around the room, while he threaded his fingers into his sandy hair. 'You are more interested in shoring up your own power here than you are in helping our Spanish brothers reconvert Northern Europe,' he accused Pope Urban.

There was a murmur of dissent from the cardinals, even from some of the other ambassadors, not because what Borgia was saying was untrue but simply because it was unwise to voice it.

'Borgia . . .' began one of the elder cardinals.

'No, I will not keep quiet. King Philip is spending his fortune making sure the Catholics of the North can press the war to their advantage. What is our own Pope doing to help? Nothing.'

'That is enough,' said a shrill voice. Pippe's eyes darted to its source. It was the Pope's nephew, Cardinal Francesco Barberini. The face beneath the tight curly hair was drawn.

Pippe caught the thrill of knowing what was about to happen. Barberini was going to break. Pippe could see it in the way the young man had his hands pressed palm down on the tabletop, arms rigid.

Sure enough, Barberini jumped to his feet, sending his chair flying.

'What about the Spanish manoeuvres in Naples?' He aimed an accusatory finger as if it were a musket barrel. 'If fighters are needed in the North, why does King Philip build up his military presence on our very borders? You forgot to tell us of this. Perhaps we have a right to be wary of Spain. Perhaps we have a right to question your loyalty.'

Now Borgia shot upwards, too. 'Of what do you accuse me?'

The two men rushed at each other, as the others stared dumbly, paralysed by shock. The doors burst open, and the Swiss Guards ran in. Pippe was impressed by their speed. He always thought of them as little more than ornaments yet in seconds their wiry frames had separated the brawlers with crossed halberds.

Breaking the stunned silence that settled over the room, the Pope spoke: 'That's enough for today.'

His voice gave the impression of control, but Pippe could hear the quaver beneath betraying temper, perhaps fright. It was rumoured that Urban lived in fear of Spanish assassins and had taken to employing food-tasters.

The cardinals and ambassadors gathered their papers and scuttled away like schoolchildren dismissed from class. Pippe dallied, undecided about whether to join them or press ahead with his plan. When he found himself the last in the room, the decision was made for him.

The Pope raised his head. 'Cardinal Pippe? What troubles you?'

'Your Holiness, it's Galileo. His new book . . .'

'His *Dialogues*? It's the most eagerly anticipated book of astronomy.'

'Although it pains me to say it, he uses the work to attack traditional thinking and . . .'

'There is nothing wrong with modernity.'

'. . . and to assert absolutely that the Earth moves.'

Pippe saw the words strike Urban.

'And how does he do that?' The Pope attempted to sound nonchalant.

'He spends a great deal of the book discussing why the ebb and flow of the tides prove conclusively that the Earth moves through space – an argument I need not remind you has been rejected as utterly false by the Roman College.'

Urban pushed himself from the throne, walked to the window and stared outwards with his hands clasped behind his white robes. Pippe held himself motionless, wondering what else he should be saying.

'Send me this book. I will have it read to me,' said Urban.

'It is five hundred pages long.'

'Five hundred pages? It would take me a feast every day for a month to get through it.' Urban turned to face the cardinal, 'I want it read by three others; they must decide Galileo's intention with this work.'

'Yes, Your Holiness.' Pippe found it difficult not to smile before he was safely out of sight.

A week later, Pippe returned to the audience chamber carrying the assessment. Urban edged forwards as he listened to the conclusion.

'We think that Galileo may have overstepped his instructions by asserting absolutely the Earth's motion and the Sun's immobility, thus deviating from hypothesis,' read Pippe carefully. 'Your Holiness, we must now consider how to proceed against Galileo and the book.'

Urban's gaze drifted from Pippe. 'How is one supposed to proceed in such a matter? When Galileo visited me, we spoke of his writing this book. Perhaps I even urged him to write it.'

'This might help.' Pippe produced another sheet of writing, recapturing Urban's attention. 'It's the edict from 1616, issued to Galileo by Cardinal Bellarmine, may he rest in peace. It clearly states that Galileo was forbidden to hold, defend or to teach, in any way, the Copernican doctrine. *In any way*, Your Holiness.' Pippe's eyes fixed upon the Pope. 'Just by discussing this book with you, he was breaking the law.'

'No, there must be some other explanation. Galileo must have misunderstood Bellarmine.'

Why was he being so stubborn? Pippe tugged his bottom lip, thinking hard. 'Your Holiness, I believe that Galileo has mocked you in his pages.'

'Ridiculous. It is a work of philosophy.'

'Your Holiness, with all respect . . .'

'Today is not the day for scheming, Cardinal Pippe. Explain yourself.'

Pippe tensed; he should have rehearsed this. He hung his head in imitation of a penitent. 'The book is a dialogue between three people. Salviati is a thinly disguised version of Galileo, spouting Copernican nonsense at every opportunity. Sagredo is supposedly neutral but sides with Salviati. Then there is the bastion of tradition and ancient reason; the spokesperson for our way of life. Painted here as an imbecile, even in name. He is called Simplicio.'

'Simplicio?'

'Your Holiness, it is my most humble opinion that Galileo has cast you in the role of Simplicio. The similarity of the name to that of simpleton is one you hardly need a Jesuit intellect to recognise. Galileo did not write the book in Latin but in Italian for the common man. He is claiming to all that he is your intellectual superior.'

Urban's eyes were unblinking. Pippe felt uncomfortable under their hardening glare and willed himself to keep talking. 'Is it not your belief that God's will is greater than any man's imagination? Therefore how a man chooses to interpret nature can never be held up as true?'

Urban nodded tightly. 'Go on.'

'Allow me to read to you the words of Simplicio, his last line of defence, once Salviati has supposedly beaten him in all other argument. He appeals to God's omnipotence: *it would be excessive boldness for anyone to limit and restrict the Divine power and wisdom to some particular fancy of his own.* Your Holiness, Galileo has placed your very philosophy in the mouth of a simpleton. Can there be a greater insult?' Pippe regretted the words as soon as they left his mouth. He had gone too far, he was sure of it. He began to feel cold and clammy.

When the Pope finally spoke, it was quietly, as if he were organising his thoughts. 'Betrayed by a man I sought to make an ally, a man whose intellect I respected. And all the while he was making sport with me. Cast me as a fool, would he?'

For a moment, Pippe thought Urban was going to cry. The Pontiff raised his hands to his head and rocked back and forth. Presently he lifted his face to stare at Pippe again; all traces of self-pity erased.

'Fetch him to Rome at once. He will answer for this.'

34
Linz, Upper Austria
1627

Wooden blocks had been placed behind the wagon's wheels, and for the last hour two industrious drivers had packed it with clothes and chattels. They had fitted everything together like some giant puzzle and topped it off with Kepler's writing desk, upturned so that its legs resembled chimneys. Now, they were making their last inspection of the ropes holding everything in place.

Kepler dodged between them, performing his own investigation. Try as he might, there was no way he could squeeze in the armillary sphere he was holding. He attempted to balance it in the far corner but realised that the first pothole would send it bouncing from the wagon. Reluctantly he retrieved it and carried it back into the house.

'We'll have to leave this one behind until we can send for the last of our things,' he said.

Susanna was crouched in the hall, fastening Fridmar's jacket. 'I never realised just how many possessions we've accumulated. Do we really need all of them?'

'I'll be sure to remind you of that the first time you ask me where something is.' Kepler set down the sphere, running his hands across its filigreed surface.

Susanna flipped Fridmar's hair from his eyes. 'Off you go, into the carriage. Your brother and sister are waiting.'

From the doorway, Kepler could see the carriage they would be travelling in, parked behind the wagon. Inside it, Cordula was swinging her legs, impatient for the adventure to begin, an arm curled around Hildebert who grinned at the people gathering in the street to watch their departure. Fridmar clambered in beside his brother and sister. Kepler was about to follow but stopped on the doorstep and took a deep breath.

'What is it?' asked his wife.

'Once more my belongings and I are bundled together for a journey. It seems to be the story of my life: never quite knowing what awaits me. If I were a planet, I'd know exactly where I was going.' He raised an eyebrow to let her know he was not being entirely serious.

'What do you hope for this time?'

Kepler ran both hands down his beard. 'I don't know. But it feels different. For the first time in my life, I have discharged all my debts to other people – all my debts save one, that is.'

'And who is that to?'

Kepler drew her close. 'To my wife and children. I owe them some years of peace and quiet.'

Susanna put her arms around him. 'I think you're about to fulfil that debt.'

'So do I.' Kepler bent his head to hers. The warmth of her lips reminded him of the first time they had kissed. It had been in the meadows downstream from the town. He had hired a boat to take her there, forgetting what hard work it would be to row back upstream at the end of the day.

They had talked and laughed and looked at one another. Then they had leaned quite naturally together. The instant before their lips touched, he remembered Barbara and felt a rush of panic. It vanished upon contact. Susanna's hot lips seared his, and he knew it was right to love her.

He lingered in the feeling and forgot about the poverty, the illnesses and the tragedies. All that mattered now was their future. He pulled back and looked into her eyes.

'To Sagan,' he said.

'To Sagan,' she repeated.

He took her by the hand and led her to the waiting carriage.

35
Rome, Papal States
1633

Galileo had taken to prowling the hallway of the Tuscan embassy. Today, however, the heat did not agree with his breathing. He propped himself against a table, massaging his aching chest.

Once a week a courier arrived to deposit a pouch containing letters and documents of state, and to remove the previous pouch now stuffed with outgoing post. It was one of those automatic processes that allowed the embassy to function. No one paid it much mind. But on this particular morning, Galileo was waiting when the courier galloped into the grounds. He watched one of the administrators accept the correspondence.

The tubby young man caught Galileo's stare. 'Are you expecting something, signor?'

'My daughter,' wheezed Galileo.

'Your mail will be brought to your room once I've sorted it.'

'Please,' said Galileo.

Pity crossed the man's face. It irked Galileo, but if it meant he got the letter faster, he would tolerate it.

The administrator made a laborious search of the pouch. Then he turned to Galileo. 'Sorry, signor.'

Galileo had been here two months now, dragging himself from room to room, staring at the walls, enduring day after day of relentless heat. Plague was spreading across the peninsula, and he spent endless hours listening to gossip about roadblocks and quarantines, of people being boarded into their homes or into inns to contain the spread of the disease.

And every day Galileo suffered the interminable wait to be summoned that final distance across the Tiber and into the Vatican. He could not decide what caused him more anxiety: the hollow relaxation of being spared the ordeal for another day, or the shallow hope that the silence meant the matter was close to being dropped.

Perhaps the real frustration was that with each passing day, another twenty-four hours of his remaining life ebbed away, wasted. He feared he would never see Maria Celeste again.

If only he could have seen her the last time he visited, but sheets of parchment had been placed over the metal grilles in the meeting room as a precaution against the plague and all he had seen was her shadow.

He had gone to her shortly after the Vatican summons arrived, inching down the hill convinced that he was on the verge of collapse. She sensed it of course, even though she could not see the way he had slumped in a chair. 'You cannot travel in your condition. I can hear in your voice how unwell you are.'

'I've tried everything. The doctors have signed an affidavit confirming my infirmity yet the Inquisition dismisses it as a delaying tactic. I'm to report to Rome at once of my own free will, or I'm to be arrested and transported as a common felon.'

'But the plague is afoot. You must not travel.'

'I must obey His Holiness. The Grand Duke is paying for me to lodge at the Tuscan embassy yet again. I'm told the new ambassador is young and energetic. My things are being packed.'

'Is Signora La Piera proving helpful?'

'She's a good housekeeper. Mind you, at seventy, I am grateful for any assistance I can get.'

'I count your birthdays at only sixty-eight.'

'Well, I feel a lot older.' Galileo reached into a pocket. 'I have something for you.' He heaved himself forwards and slid a folded sheet of paper between the translucent screen and the crumbling plaster.

'It's my last will and testament,' he said.

The shadow drew closer to the parchment. 'Father, I beseech you not to grasp the knife of these current troubles by the sharp edge. If you do so, it will only cut you more deeply.'

'I keep all your letters, you know.'

'Father, wait.' Her voice was anxious. 'There is something I must get for you. It will only take a moment.'

Galileo leaned back in the chair; its ancient wood creaking beneath him. His head was fuzzy and he tried not to think about the climb back up the hill.

The shadow returned. Galileo grasped the pellet-shaped object that was slipped under the parchment screen. It was a bottle of transparent fluid.

'It is the healing water from Abbess Ursula of Pistoia. It will help ward off the pestilence.'

Galileo admired the golden ribbon that adorned its neck but did not know how to respond.

Maria Celeste said, 'There's something else.'

A second shadow fell upon the screen.

'Father?' It was not Maria Celeste's voice.

'Arcangela, my child,' Galileo's voice caught in his throat.

'Yes.' The word sounded as if it were the prelude to a conversation but further words did not come from her.

'Are you well?' asked Galileo.

'I am.'

His memory filled with pictures of her as a child, small and neat, possessed of boundless energy and constantly laughing. She had been quite unlike her studious sister. What had possessed her to become so withdrawn? It troubled Galileo and had become one of his midnight worries. Something inside him needed to apologise to her, yet for the life of him he found the urge unfathomable. Perhaps such apologies were all that was left to a parent when their child's life had turned miserable for no apparent reason.

There was some urgent whispering, and then Arcangela spoke again. 'I will pray for you on your journey, Father.'

Now Galileo was glad of the parchment: it hid his tears.

The noise of new arrivals drew his attention back to the porch. Servants came running as a shadow fell across the entrance. There was a crunch of gravel and the creaking of a carriage coming to rest. A horse whinnied and stamped its foot.

Ambassador Niccolini, his forehead shiny with sweat, swept into the embassy. He saw Galileo at once. 'I have news. Come to my office.' He turned to the servants without breaking stride. 'Fetch us wine.'

Inside the panelled office, Niccolini peeled off his jacket revealing a skinny frame. Three of him could have fitted into the space taken up by Galileo.

A servant rattled a tray of drinks onto the table and crept quietly from the room. Niccolini faced Galileo. 'The waiting is over. The Holy Office is ready to proceed against you. You are to be moved today to the Inquisition Palace, perhaps even questioned today. Afterwards you will be detained there until this business is settled.'

'Imprisoned?'

'Not in the dungeons. You will be housed in a state room.'

'Did you show them the Duke's letter?'

Niccolini hesitated.

'You didn't, did you? What gives you the right to deny me my defence?' It confirmed everything Galileo had suspected; Niccolini was too young for this office.

The ambassador loosened the chemise laces at his neck and drank deeply from his glass. 'I assure you, it would have done no good. The Holy Father's mind is set firmly against you. It could only have damaged the Duke if I'd presented the letter and I can't believe you would have wanted that.'

'So, I am to be sacrificed.'

'You have stirred such passions. You do have allies, but your defence must be your own.'

'Who still supports me?'

Niccolini regarded him with calculation. 'Better for your cause if you do not know.'

Once more there was a crunch of hooves on the gravel outside. Niccolini's head shot round at the sound. Galileo's insides turned to ice. He pushed himself to his feet and straightened his tunic top. As the Vatican envoys escorted him from the room, he realised that he was not giddy any more. His mind was clear and focused.

The room he was taken to in the Vatican was smaller than he expected. Just three people waited inside: a secretary, here to transcribe the interview, and two clerics sitting on a small dais. One was sharp-faced with a Roman nose, perhaps not much younger than Galileo. Blue eyes stared from either side of the prominent bridge. 'I am Fra Vencenzo Maculano da Firenzuola.' His boots peeped out from beneath his flowing robes. They were polished so brightly they reflected the room.

'Father Maculano.' Galileo bowed his head.

'This is Cardinal Pippe,' Maculano indicated the man on his left.

Pippe had a blunt face with a high cliff of a forehead. He looked as if he had swallowed his upper lip and his chin was heavily dimpled. Galileo disliked him on sight. Neither acknowledged the other.

Maculano cleared his throat. 'Now then, Signor Galileo, do you know or perhaps can you guess at the reason for your presence here today?'

'I imagine that it is on account of my recently published book.' Galileo kept his voice neutral.

'What about the book requires your presence here?'

'It is a dialogue about the two systems for understanding motion in the heavens.'

'Would you recognise your book if it were shown to you?' asked Pippe.

Galileo favoured him with a long look. 'I would hope so.'

Pippe handed Galileo a copy. 'Do you acknowledge this is your book? That every word in it is your own?'

What a farce, thought Galileo. This confirmed everything he had heard about the Holy Office since Bellarmine's death. *What hope for Catholicism with these people policing it?*

He returned the book and lifted his eyes to trace the line where the ceiling joined the wall. He could sense Pippe's impatience, so delayed answering for as long as he dared. 'Yes,' he said finally.

Maculano fired another question. 'Have you been to Rome before, in 1616, perhaps? If so, what was the occasion for the visit?'

Galileo spoke at once, with confidence in his voice. 'I was in Rome during 1616 to clarify certain points about the opinions of Nicolaus Copernicus, in order to assure myself that I was not holding anything but holy and Catholic opinions. So, I came to hear what was the proper opinion to hold on the matter.'

'And what is that opinion?' asked Pippe.

'That it is repugnant to Holy Scripture and is to be taken only as a hypothesis, in the way that Copernicus does.'

'Who notified you of this?'

'The Lord Cardinal Bellarmine'

Pippe glanced at Maculano with barely disguised smugness.

'We have a written version of that injunction.' Maculano dangled a yellowed sheet of parchment. 'It states that you were forbidden to hold, defend, or teach the said opinion in any way whatever.'

A written injunction! Galileo was not expecting this. Bellarmine had never mentioned that it was to be placed in writing. The cardinal had only met with Galileo in the cloisters to tell him of the outcome. Galileo hid his surprise with a shrug. Now was not the time to deviate from his strategy. 'I do not remember being given this in anything other than verbal form. I remember Lord Bellarmine informed me that I could not hold or defend – maybe even that I could not teach –

but I do not recall that the phrase "in any way whatever" was used.'

Maculano raised his eyebrows. 'Did you obtain any permissions to write your latest book?'

'I did not seek permission to write my book because I did not think that I was contradicting the injunction.'

'Really?' Maculano let the slightest hint of boredom enter his voice.

Galileo seized on it. This was just a formality: they would ask him their tiresome questions; he would give them a few pat answers. Then his testimony would be filed away and forgotten. Why else all the prevaricating in calling him to the Holy Office? It was clear to Galileo now; he just needed to say the right thing and he would be on his way home.

'The edict stated that I was not to hold, defend, or teach the said opinion – and I wasn't. Rather, I was refuting it in my book.'

'You were refuting Copernicus?'

'Utterly. I have not defended the view of Copernicus in any way. Rather I have shown the opposite, that the Copernican opinion is weak and inconclusive.'

There, thought Galileo, *it is said. A lie to serve a higher purpose; someone has to save these people from themselves.*

Maculano turned his head slightly but continued to look at Galileo. Pippe remained seated, his expression hidden behind a hand.

'We have him now. He has lied to us, demonstrably so. Perjury!' With so much energy running through him, Pippe found it difficult to contain his gesticulations. He paced in front of Grienberger's desk, his thoughts turning to Bellarmine. The calmness with which his old mentor had approached the Giordano Bruno case had bordered on prevarication, and Pippe was determined not to make that mistake with Galileo. 'What more evidence do we need? Why the indecision?'

Maculano frowned at Pippe and turned to the elderly Jesuit behind the desk. 'What do the reports say, Father Grienberger?'

Grienberger's hair was now entirely white and he had cultivated the jaw-line beard favoured by Clavius, his predecessor as the head of the Roman College. Age had set fast his hangdog expression. He shuffled some papers in front of him. 'I have the reports here. We have again examined the book of Galileo and we have again found that it teaches the Copernican error of Earth's motion through space. This is now beyond all doubt in our eyes. Being written in a book also signals to us the desire to teach this work not just to the present generation but to future ones too.' Grienberger inclined his head with an air of disappointment. 'Had he intended to extend his ideas to learned men of different nationalities for discussion, he would have written in Latin. Yet he writes in Italian to entice the common people who sadly lack the education to judge his ideas, and in whom errors can so easily take root.'

Maculano wrung his hands. 'Why does he deny the obvious? What can he achieve by treating us as fools?'

'He's trying to save his own skin,' said Pippe. 'He treats us with contempt and must be punished accordingly.'

'Galileo has his faults, but I can't think of him as anything other than a basically honest Catholic,' said Grienberger. 'If he'd truly been intent on harming the Church, he would have stolen to France or Germany and published his book there. He didn't do that. Instead, he seems to have a personal desire for us to embrace Copernicanism.'

'He's a liar, nothing more,' said Pippe. 'If Galileo continues down this path, we'll have no choice but to burn him.'

Despite the summer heat, Galileo felt cold, even shivery. Confined to the apartment, he was transparent to the warmth in which Rome basked. He pulled on a fur-trimmed cloak and stood at the window. The sunlight glinted from the gold of the statues, mocking him to come outside. He was surrounded by comfort, but there was nothing to occupy his mind. So he stood and stared at the sculpted figures. As the hours crept by, so each shadow inched around revealing the rotation of the Earth.

He watched the people on the approach to the Vatican, each tiny figure dragging around its own shadow. They were the closest he came to human contact on most days.

There was a soft knock on the door. Galileo was not expecting Maculano to be standing in the corridor, yet it was the Father's pointed features that greeted him when he pulled open the heavy door.

The visitor looked uncomfortable and hurried into the room, urging Galileo to shut the door quickly. He took a seat but refused a drink. 'Galileo, I am here to reason with you.'

'Reason with me?'

'Yes, to persuade you to change your course.'

'But I'm the victim here. I'm persecuted because of my enemies' vicious plots against me.'

'There is no one who will defend your position. Even those who applaud your efforts state the book is pro-Copernican.'

'In truth, Father,' Galileo shifted some pleading into his voice, 'I have done nothing to defend Coper—'

Maculano's hand sliced the air. His voice rose in volume. 'Drop this fiction! Each time I send the book out for review, I hope for some small nugget of doubt in your intentions, and each time the condemnations come back stronger than before. If you'd attacked an individual thinker over the inadequacy of a personal argument in favour of a stationary Earth, we might have been able to put some favourable aspect on this work. But, no, you declared war on everyone who was not a Copernican. Your position is indefensible.'

Galileo sat down, resting his hands on his lap. They were pale and bony, with swollen joints. And they trembled.

Maculano spoke again, his voice quiet. 'I'm not insensitive to the virtues of Copernican astronomy. Really I would prefer that matters of astronomy be separated from holy law, but that is not how it is, or perhaps how it will ever be.'

Galileo looked up at Maculano. Those blue eyes, which had looked so icy in the courtroom, were pleading with him now. 'Signor, I am to question you again tomorrow. I pray you reconsider your position. If you do so, the tribunal can retain its reputation and deal with you benignly. I need not tell you what could lie in store if you persist in your denials. The evidence against you is overwhelming.'

Galileo's throat tightened. He swallowed but found no relief. That night, he dreamed of flames.

Galileo found it difficult to stand in the hearing room the next day. The tremors that had begun in his hands invaded his legs. On the walk over, the Swiss Guards had supported him. Now they had withdrawn, he felt vulnerable. His beard caught in the fur of his cloak as he looked around. The movement brought on a fit of giddiness. Rubbing his temples, he felt the blood pounding through his head.

'Signor Galileo, is there anything you would like to say?' Maculano's voice sounded distant, and the room was too bright.

Galileo ran his tongue around his lips, moved his fingers to smear the dampness around, then forced himself to speak. 'For several days I have been thinking about the previous interrogation. In particular, about the question of whether – sixteen years ago – I had been prohibited by the Holy Office from holding, defending, and teaching in any way whatsoever the opinion, then condemned, of the Earth's motion and Sun's stability.'

Galileo sensed a different version of himself speaking, divorced from his true self. The situation proved both unnerving and curiously liberating. Inside, the true Galileo clung to the truth. Surrounding this kernel was the altered Galileo that everyone else saw.

'It dawned on me to reread my printed *Dialogue*, which I have not looked at for three years. I wanted to check very carefully whether, against my purest intention, there might have fallen from my pen not only something enabling readers – or superiors – to infer disobedience on my part, but also other details through which one might think of me as a transgressor of the orders of Holy Church.'

It occurred to Galileo that there was something truly wicked at large in the universe. Until yesterday he had only dimly perceived it, and thought that the problem lay in man's perception of the heavens. He had fully believed that he would be able to fix it. Now, he understood that the maleficence was so strong that he had been a fool to attempt its excision.

The altered Galileo spoke again. 'Not having seen it for so long, I found it almost a new book by another author. Now, I freely confess that it appeared to me in several places to be written in such a way that the arguments for the false side were stated too strongly. In particular, two arguments, one based on sunspots and the other on the tides, are presented as being more powerful than would seem proper for someone who deemed them to be inconclusive and wanted to refute them – as indeed I inwardly and truly did. Having fallen into an error so foreign to my intention, to use Cicero's words, "I am more desirous of glory than is suitable." If I had to write out the same arguments now, there is no doubt I would weaken them. My error then was, and I confess it, one of vain ambition, pure ignorance and inadvertence.'

A wave of exhaustion overwhelmed him. He felt himself sway and tried to lock his muscles in an attempt to stay upright. 'This is as much as I need to say on this occasion.'

Maculano was nodding. 'Thank you, Galileo. I see no reason to detain you at the palace. You are free to return to the Tuscan embassy but not yet to leave Rome.'

As he turned, Galileo caught sight of the other cardinal, the flat-faced one, scowling. It perturbed him, even as he was ushered away down the corridor, resting on the guards again. The cardinal should have been happy with what he had said. It was what they wanted him to say, was it not?

Confusion crowded his brain. Why was Pippe – yes, Pippe, that was his name – scowling? Was the confession not good enough? Perhaps he should have said more.

Galileo turned. The guards caught him by the shoulder as he began to fall.

'Please, I must speak again,' said Galileo, planting his feet.

Without a word, the guards took him back to the room, where Maculano and Pippe were preparing to leave.

'I could write another one or two days in the book, so that the speakers agree to meet again. I promise to reconsider the arguments already presented in favour of the condemned opinion and refute them in the most effective way that the blessed God will enable me. I beg this Holy Tribunal to grant me the permission to put it into practice.'

'The last thing we need is you writing another book, Galileo,' said Maculano. 'Go and rest.'

'Sir, sir, there's a letter for you.'

Galileo peered into the gloom. He pushed himself upright, his shoulders stiff and cold as ice.

The man thrust the letter in front of him. 'I believe it's the one you've been waiting for. From Florence. Shall I read it for you, sir?'

Galileo shook his head. 'No, that will be all. Thank you, Tito.'

When the door closed, Galileo fumbled open the letter. He recognised the handwriting instantly, though it was smaller and tighter than usual. He skimmed the letter so fast, he only picked up the basic themes: Arcangela in trouble, a request for money, the plague raging and finally a plea for him to drink less.

Hmm. So what if he drank in the evenings? It had been a mistake writing to Maria Celeste with an enthusiastic description of the vintages held in the Embassy's cellar. But what else was there to tell her about? There was nothing else to do.

He was forbidden social visitors. While the Embassy was indeed comfortable, he might just as well be in prison. No one had time for him here, apart from Tito. They were all so busy with their jobs that Galileo was nothing but a ghost, haunting the corridors. Since submitting his written defence, he had done nothing.

As he reread the letter more slowly, he realised that his return to Florence was not going to happen soon. The city was being ravaged by the plague, so much so that the Grand Duke had ordered the sacred painting of the Madonna to be taken from the church in the nearby village of Impruneta and paraded through the Florentine streets to drive out the evil of the plague.

How Galileo wished he could have seen the giant portrait being carried aloft. Each night the icon was being sheltered in a different religious building. The nuns at San Matteo had broken down a wall because their unobtrusive doorway was too small to grant it passage. Galileo could imagine Maria Celeste's hands clearing the rubble to offer the painting sanctuary. Thankfully the plague was still clear of the convent, but it gripped the city so tightly it would be madness to return until the autumn, months away, when it would naturally abate in the cooler weather.

He read on. Arcangela was in debt. She had overspent on the provisions for the convent and needed funds. He would have to write at once to organise a release of money. Ordinarily Galileo would have bristled at his younger daughter's incompetence; today he relished having something to do.

Maria Celeste next recounted that she had successfully exchanged Arcangela's spell of looking after the wine cellar to looking after the convent drapery. Apparently Maria Celeste worried that Arcangela might drink too much if she were appointed Cellarer. And that led her neatly into nagging her father about his own wine consumption. He read her soft admonishment with the whisper of a smile on his lips.

He rose from the bed and went in search of pen and paper. For the next hour he could be busy and feel worthwhile again. All too soon he would return to the endless waiting for the verdict. Why was it taking so long? They had what they wanted from him. Why not just publish his humiliating retraction and let him go? No one would take him seriously any more. What more harm could he do?

The cardinals took their seats in the traditional horseshoe formation around Urban. It crossed Pippe's mind that the Pope was looking beleaguered these days. Like so many of them now, Urban's hair had turned to snow, as had his beard and moustache. Only his eyebrows retained their dark colour and, together with Urban's dark eyes, they radiated gentility.

Perhaps a little too much gentility, thought Pippe. *We need a strong Pope, especially at times like this.*

Urban's forehead was riven with worry lines. As the cardinals debated, he would frequently massage the bridge of his nose, or press at the tight knots of flesh just above. He would look out at them with a pained expression, then drop his gaze to the floor.

'What are we if we destroy this man?' Barberini, the Pope's nephew, was saying. 'What purpose will it serve?'

'We must be strong in our defence of Catholicism,' answered another.

'You were a student of Galileo's, were you not?' Pippe asked Barberini.

'My education does not enter into this.'

Pippe held his gaze.

'Yes, I was tutored by Galileo,' admitted Barberini. 'I knew him well and continue to regard him highly.'

A frisson rippled around the chamber.

'That does not mean I believe Galileo is innocent of all charges,' said Barberini.

'He is a heretic and there is only one punishment for heretics,' said Pippe.

'Careful.' Urban wagged his finger at Pippe. 'The 1616 decree does not include the word heresy. In fact, I remember that the word was specifically removed. Whatever Galileo is guilty of, it is not heresy.'

'Then he is vehemently suspected of it,' said Pippe, registering the ripple of agreement around the table. 'Suspicion hangs around him like the smell of drains in summer. Your Eminence, this is no time for leniency. He persists in the lie that he never held a belief in Copernicus.'

Urban turned his head. 'Is this true, Father Maculano?'

Maculano shot Pippe a look. 'It is true that Galileo claims not to have believed in the Sun's stability following 1616, and that he claims his *Dialogue* was designed to falsify Copernicus, not prove it.'

Urban pinched his nose and closed his eyes.

Pippe spoke again. 'Lying to the Inquisition is a most serious affair. We cannot be lenient.'

'I will not make a martyr out of Galileo,' said Urban.

'He makes fools of us all.'

Urban pressed his palms together, raised his fingers to touch his lips. 'There is no doubt in my mind that Galileo is lying about his belief in Copernicus. He must therefore be severely punished and may God forgive him for being so deluded as to involve himself in such matters in the first place. But before Galileo's punishment, he must be given one last chance to confess the true motive for writing the book. He is to be questioned under threat of torture but he is not to be taken to the dungeons, nor shown the instruments. Do you understand?'

Pippe was on the verge of protesting when Maculano spoke. 'Perfectly, Your Holiness.'

'Once we have his final deposition, we can decide his eventual fate,' said Urban.

'Your Holiness, he's an old man. Has he not suffered enough?' asked Barberini.

'His age makes our actions all the more urgent,' said Urban. 'One must meet God cleansed of sin. If Galileo will not cleanse himself willingly, then we are forced to do it for him.'

Pippe smiled at the prospect.

'Once more, Galileo, we are here to establish your true intention in writing the *Dialogue*.' Maculano sounded bored, but this time Galileo knew better than to read anything into it. What was beyond doubt, however, was that he was here, with the same two cardinals for the fourth time. That had to mean something.

If only he could force his mind to work. He had offered to correct the *Dialogue*, why was that not enough?

Maculano spoke again. 'Do you have anything to say?'

That was when it hit Galileo, as it used to when he was young: the sudden clarity of thought. Felt more than seen, everything suddenly made sense. Now he knew the truth. He could see it on the two cardinals' faces: the book was to be banned altogether. He was to be denounced as a heinous criminal.

What now for Galileo? What now?

He forced himself to calm down. 'I have nothing to say.'

'How long have you held the Copernican arrangement of Heaven to be correct?'

'A long time ago, before the decision of the Holy Office, I was . . . undecided regarding the opinions of Copernicus. It could be true in nature. But after the said decision, I never held it. The Earth's stability is true and indisputable.'

'Yet you are presumed to hold the opinion because of the manner in which you presented and defended it in your book. I ask you, therefore, to freely tell me the truth as to whether you hold or have held the opinion.' Impatience had seeped into Maculano's voice.

Why won't you believe me? How many more times do I have to say it? This is the truth now, the one I feel in my heart. The one I wish to drape across my past. Surely penitence is all that matters now?

'I did not write the *Dialogue* because I held the Copernican doctrine to be true. I tried to show that neither has the force of conclusive demonstration and so we have to proceed with the Holy Father's guidance.'

'Galileo,' Maculano's voice was now grave, 'unless you proffer the truth, we will have no choice but to rely on torture. Do you understand me?'

Pippe was leering from his seat.

A wave of nausea swept over Galileo. He had heard of the pricking needles, the branding irons and the stretching racks. He gagged at the mere thought; his body hurt enough from the decrepitude of old age. That was sufficient punishment alone.

How could you be so wicked to one of your own, a loyal servant? It was beyond reason. Confess and burn, or lie and be tortured. Either way he was lost. Galileo knew he could not survive the torture chamber. He would never see Maria Celeste again. *I'm so sorry*, he said to her silently. Then he began to weep.

He hated himself for each solitary tear that slipped across his cheek and gathered in his beard. He took a final deep breath. 'I am here to obey, but I have not held this opinion after the determination of the Holy Office. You must do with me as you please.'

There was a long pause then Maculano rose from the chair. Galileo fought to remain upright. A cyclone of confusion buffeted him as he waited for the command that would see him dragged to the dungeons, and the pricking needles.

'You may return to the Tuscan embassy . . .'

'No, wait,' said Pippe, shooting from his chair.

'Enough!'

It's too perplexing. Galileo listened with only vague comprehension. His ordeal seemed to be over, he was to await sentence.

Several days later, Galileo trudged beneath the gothic arches in the Church of Santa Maria Sopra Minerva. Spared the journey to the Vatican, he had been called only as far as the city centre. Initially he had been grateful but when he saw the nave's ceiling it touched raw nerves. The very setting was a taunt. The domed arches were painted navy blue and studded with golden stars.

He changed into the set of pure white vestments handed to him and followed the guards to a spiral staircase. He climbed slowly, tripping on his robe because he found it difficult to hold the rail and prevent the material from wrapping around his feet.

When he reached his destination, he saw the room was filled with cardinals and other officials. An unnatural hush fell across them as he appeared. He paused at the door to look at his accusers. Even Grienberger was here, characteristically avoiding eye contact. Galileo stepped inside. He identified Maculano and shuffled across to stand in front of him, head bowed.

Maculano read from a sheet of parchment. His voice was strong and betrayed no emotion. 'In the judgement of this Holy Office, you have rendered yourself vehemently suspected of heresy, namely in having held and believed the doctrine which is false and contrary to the Sacred and Divine Scriptures, that the Sun is the centre of the universe and that the Earth moves. Consequently you have incurred all the censures and penalties enjoined and promulgated by the sacred Canons and all particular and general laws against such delinquents.'

The room spun around Galileo. That could only mean burning, he was sure of it.

Maculano continued. 'We are willing to absolve you from them provided that you, with a sincere heart and unfeigned faith, in our presence abjure the said errors in a manner that we will prescribe to you. Furthermore, so that this grievous and pernicious error does not go altogether unpunished, we order that the book *Dialogue* be prohibited by public edict.'

Maculano looked around before completing his statement. 'Also, as a salutary penance, we impose on you to recite the seven penitential psalms once a week for the next three years.

And, finally, we condemn you to formal imprisonment in this Holy Office at our pleasure.'

Imprisonment instead of burning. A life but no life.

Before he had time to think more, he was presented with a sheet of parchment. The curly script contained his recantation. He read it slowly; taking pleasure in letting the assembled men wait. When he finished the text, he went back to the top and read it again. Then, he handed it back and turned to Maculano.

'I will not speak these words.'

A pained expression materialised on Maculano's face. 'Don't you know what you risk? Why ever not?'

'It states that I lapsed in my adherence to Catholicism. I deny this. My personal judgement may have lapsed but I have never lost my love for the Catholic Church, not even now. I will not read a passage that states I am a bad Catholic.'

There were murmurings. Maculano drew a number of the cardinals together and a hushed debate ensued. The result was that Maculano motioned to the scribe who crossed through the disputed words.

'You will now recant,' Maculano said to Galileo.

Two guards supported Galileo and lowered him to his knees. He was passed a copy of the Bible and the script. It quivered in his hand as he raised it to eye level. The room fell silent.

Once more, he told himself, and began reading aloud. 'I, Galileo, son of Vincenzio Galilei, Florentine, aged seventy, arraigned personally before this tribunal against heretical depravity, believe all that is held, preached and taught by the Holy Catholic Church. For the act of printing a book that argued in favour of an already condemned doctrine, which is that I believed the Sun is immovable and the Earth moves, I have been judged vehemently suspected of heresy.'

Galileo paused for breath. 'Therefore, desiring to remove from the minds of your Eminences, and of all faithful Christians, this vehement suspicion, justly conceived against me, with sincere heart and unfeigned faith I abjure, curse and detest the aforesaid errors and heresies, and generally every other error, heresy and sect whatsoever contrary to the said Holy Church. I swear that in the future I will never again say or assert, verbally or in writing, anything that might furnish occasion for a similar suspicion regarding me.'

Galileo paused again. It was so hot in the room he could feel the sweat beading on his forehead.

'Finish the reading.' It was the voice of Pippe.

Galileo unhurriedly mopped his brow then returned his eyes to the text.

'In the event of my contravening, which God forbid, any of these promises and oaths, I submit myself to all the pains and penalties imposed and promulgated in the sacred canons and other constitutions, general and particular, against such delinquents. So help me God and these, His Holy Gospels, which I touch with my hands.'

He looked up, not needing to read the final line, the hollowest of them all. He recited, 'I, Galileo Galilei, have abjured as above with my own hand.'

36

The housemaid looked around, surprised at the speed with which she had finished Galileo's packing. She was olive-skinned and dark-eyed, young as well, though the details of her features were beyond Galileo's ability to discern.

'I have nothing, I am nothing,' he said, leaning on the stick he had taken to using. 'Where am I going?'

'I don't know, sir.'

'You must know. Why won't anybody tell me anything?' He tried to watch her move around the room, but she was too quick for him, and he became light-headed. 'I want to stay here.'

'The other day, you told me you wanted to go.'

Galileo could not remember the occasion. 'Well, now I want to stay. I'm too ill to travel – oh, do stand still – I shall die if I'm moved. You'll have me on your conscience.'

'Come, Galileo, you're strong as an ox.' The voice was male, if lacking a certain timbre.

Galileo turned to see Niccolini. The Tuscan ambassador was dressed informally in a linen tunic.

'That will be all, thank you,' said Niccolini to the housemaid, who flew from the room.

'Where am I going?' Galileo demanded.

'We have struck a bargain with the Inquisition. You are to escape imprisonment if you remain under house arrest.'

'For how long?'

'For the rest of your life, Galileo.'

Galileo pretended not to hear.

Niccolini continued. 'The Archbishop of Siena, Ascanio Piccolomini, has agreed to take you in . . .'

'Why can't I go back to Florence?'

'. . . He counts two Popes in his lineage.'

'Why can't I go back to Florence?'

'You cannot go back to your old life, Galileo. Your sentence can only be fulfilled by a custodian acceptable to the Inquisition. Archbishop Piccolomini is such a man. Cardinal Barberini was most helpful in arguing for your wellbeing.'

Galileo wondered again if by ignoring him he could just pretend nothing had ever happened.

'Come, Galileo. This is the best we can do for you. It's this or the dungeons. A carriage is waiting to take you to Siena. At least you're returning to Tuscany.'

Galileo did not move until he felt a hand take him by the elbow; the ambassador was surprisingly strong. As Niccolini guided him through the villa, Galileo noticed someone familiar in the hallway and stopped dead.

Galileo stared at the familiar face but could not think of anything to say.

'No mail today, sir. I'll be sure to forward it, when it arrives,' said Tito.

Everything seemed so meaningless. It was such a long way to the carriage and then such a long way up into the box. The ambassador lifted him inside.

Throwing out the rubbish, thought Galileo.

A short while later, the wooden vehicle rattled across the gravel and out of the embassy gates. Tito watched from the steps with the ambassador at his side, both reluctant to turn away. Before the carriage disappeared completely from view, it shimmered in the heat haze.

'So this is how it ends,' said Tito.

Niccolini leaned close. 'No one knows this – not the Inquisition, not the Jesuits, not even Galileo – but Archbishop Piccolomini is a Copernican.'

Tito looked round, convinced he had misheard. Niccolini smiled at his evident confusion.

'You mean . . .' began Tito.

'I mean it isn't over,' said Niccolini.

Epilogue

Johannes Kepler's textbook *Epitome astronomia Copernicanae*, Epitome of Copernican astronomy, became staggeringly influential. The title was a mark of his humble nature as he did not advocate Copernicus but rather his own system of the planets. Even today we talk of the Copernican model, not the Keplerian one. Yet all Kepler took from Copernicus was that the Sun was stationary and the Earth moved. Although astronomers could not prove the Earth moved until 1725, the simplicity of Kepler's elliptical orbits captured the imagination of natural philosophers across Europe and convinced them that the universe was understandable to humans. However, Kepler did not live to see this, dying in 1630 at the age of fifty-eight.

Galileo Galilei was all but broken by his trial. He was protected and nurtured by his supporters, who made sure that his house arrest was conducted in sympathisers' homes. They coaxed from him another book, *Discorsi e dimostrazioni matematiche*, Discourses and demonstrations on two new sciences, about motion and material strength. It was more influential than his astronomical works and is now acknowledged as the first work of modern physics, presenting scientific arguments in precise mathematical detail rather than rhetorical flourishes. The work was smuggled out of Italy and published in Holland.

Kepler's *Epitome* and Galileo's *Discorsi* became the foundation stones for the scientific revolution. They provided Isaac Newton with the raw material to develop his groundbreaking theory of universal gravitation, published in *Philosophiae*

Naturalis Principia Mathematica, Mathematical Principles of Natural Philosophy, in 1687.

Galileo died, aged seventy-seven, in 1642, the same year that Isaac Newton was born.

Acknowledgements

This is a story based on truth. As such, bringing it to life would have been impossible without the existing manuscripts of the astronomers involved and rendered far more difficult without the extraordinary efforts of the numerous historians and writers who have previously published non-fiction accounts of these various stories and people.

There are some wonderful biographical examinations of these characters in print, and if I have piqued your curiosity about the characters in this book at all, then I encourage you to progress to these other books, and decide for yourself whether you agree with my interpretation of events and personalities.

Two books particularly spring to mind because they not only paint the people so well but also the times in which they lived: Dava Sobel's *Galileo's Daughter* and James A. Conner's *Kepler's Witch*. Other sources of mine include *Kepler* by Max Caspar, *On Tycho's Island* by John Robert Christianson, *The Sleepwalkers* by Arthur Koestler, *Galileo, Bellarmine and the Bible* by Richard J. Blackwell, *The First Copernican* by Dennis Danielson, *The Mercurial Emperor: The Magic Circle of Rudolf II in Renaissance Prague* by Peter Marshall, *Copernicus and His Successors* by Edward Rosen, *Astronomies and Cultures in Early Medieval Europe* by Stephen C. McCluskey, *Tycho Brahe: A Picture of Scientific Life and Work in the Sixteenth Century* and *A History of Astronomy from Thales to Kepler* by J.L.E. Dreyer, *Science and Civic Life in the Italian Renaissance* by Eugenio Garin and *The Galileo Affair*, edited by Maurice A. Finocchiaro.

Then, of course, there are the books by the great men themselves: *Dialogue Concerning the Two Chief World Systems* by Galileo

Galilei, *Mysterium Cosmigraphicum*, *Astronomia Nova*, *Harmonices Mundi*, *Somnium* and *Epitome Astronomia Copernicanae* by Johannes Kepler.

I have made a few, hopefully acceptable, changes to the chronology in order to fashion this story into fiction but in spirit I believe I have remained true to the people, the science and the events. There is only one main character in this book who is entirely fictitious. That character is Pippe.

Then there are my heartfelt thanks to the people who have believed in this project and been involved in bringing it to fruition: Peter Tallack, Duran Kim, Neville Moir, Caroline Oakley, Hamish Macaskill, Maria White, Alison Rae, Jan Rutherford, Brenda Conway, Alison Boyle, Nic Cheetham, Ruth Seeley and Kim McArthur.

And of course, Nicola Clark, my wife and invaluable assistant.

COMING SOON!

The Sensorium of God

The second book in The Sky's Dark Labyrinth trilogy by Stuart Clark

Read on . . .

It is the middle of the seventeenth century. In England, there is widespread mistrust of the Stuart royalty, with many fearing their Catholic sympathies. Scientific investigation has nucleated into an elite London organisation called the Royal Society, patronised by King Charles II. The monarch has also built the Royal Observatory in Greenwich to produce a set of star tables so precise they will help navigation and make the English fleet invincible.

Robert Hooke is the quintessential Restoration experimentalist but lacks the mathematical skill to fully exploit his insights. Once at the heart of the Royal Society, his argumentative manner has lost him support. Now in failing health, he is turning bitter. His only comfort is the love he shares with his niece.

He feels sidestepped when Edmond Halley, a young adventurer and astronomer with a reputation for the ladies, visits the shadowy Cambridge figure of Isaac Newton, a reclusive alchemist with a fearsome talent for mathematics. No one understands why the planets move as Kepler described so beautifully almost a century earlier, and Halley asks Newton for help in solving the problem. Little does either know that this simple question will plunge both their lives into crisis, place Europe headlong on course for the Age of Enlightenment and catapult science into its next decisive clash with religion.

When Isaac Newton succeeds in answering Halley's question, a bold new understanding of nature opens before him. Not only can Newton explain planetary motion, he can use his concept of gravity to interpret Galileo's work on falling objects. The success of his theory emboldens Newton. He begins to think that he has been blessed with a special confidence and that God is working through him to produce a new gospel for a modern age.

But Newton has dark secrets and forbidden passions. He believes in a religious heresy so profound that he lives in constant fear of discovery. At the height of his fame and power, Newton meets a precocious Swiss mathematician who wins his confidence but the young man holds dangerously revolutionary thoughts. The secrets Newton confides in him could be just the weapon he needs to strike against the monarchy.

Halley also finds himself on a collision course with the Crown when his father witnesses an illegal royal scheme to summarily execute a plotter at the Tower of London. Having previously accepted patronage from Charles II, Halley's worst fears are realised when an investigation begins and his father disappears.

5 November 1679

London, England

It was the fifth of November, and London was in flames. Orange tongues twisted into the night from bonfires built on street crossings and patches of green. Cinders drifted upwards into the chilly air like freed souls racing to heaven. The people were supposedly commemorating God's deliverance of James I from the Catholic plot to blow up Parliament back in 1605 but to Edmond Halley something darker was permeating the revelry.

He tried to dismiss the thought as prejudice. He had been just ten years old when the Great Fire of London had raged across the city – a tragedy sparked by Catholic brigands according to some – and, even now, unpleasant memories kindled. Halley recalled being bundled into the night, unsure yet strangely excited by the atmosphere. He had been entrusted with a bundle of clothing and relied upon to walk alone while his mother carried his younger sister and gripped his little brother's hand. His father had led the way, laden with a chest of hastily packed possessions.

Swept along in a tide of people, Halley's nervousness had resolved into a sense of duty. Although there were noises all around – the occasional shout or sob, the barking of a dog or the whinny of a horse – he snatched only momentary glances in their direction. Mostly, he concentrated on marching in step behind the broad expanse of his father's back, determined to keep up.

When the family squeezed itself past an empty cart being pushed towards them, into the city, its owner made a gruff offer to carry them to safety. Halley's father curtly shook his head. A panic-stricken man carrying a frail old woman rushed in to ask the price.

'Twenty pounds to Moorfields.'

Two hundred times the normal price! thought the boy.

Above them there was a roar of thunder and orange sunbursts as first one roof then another caught fire. Cries of alarm went up as the flames took hold.

'Let's hurry up. No time to dawdle today.' Halley's father winked at him over his shoulder, and Halley found himself smiling back, insulated from the panic.

Ahead of them a stuttering Frenchman backed away from an irate mob. Halley caught a few words of his protestations and realised that he was pleading for his life. The mob were screaming that if the man was French then he must be Catholic, and that meant he was guilty of starting fires, but Halley had seen the way the wind was urging cinders through the night air and dropping them on the thatched roofs all around. New fires were sparking everywhere. He tried to speak up but his small voice was lost in the baying of the crowd. A plump, wild-eyed woman landed the first blow on the Frenchman's chin, and as the rest joined in Halley's mother ushered him firmly onwards.

On the frosty slopes of Moorfields, thousands of people had camped out. Halley could still recall the horrified look of a young woman as she stared back at the dying city, her face illuminated by its angry light. She had conjured the strangest of feelings inside him – totally incomprehensible at the time. It had seemed utterly wrong that a girl so beautiful should look so sad.

'It'll be alright,' he had said to his mother, reaching back to touch her thigh.

However, it had not been alright. His younger self had no comprehension of how precious and temporary life could be. In just a few years, his mother, his brother and his sister would be gone for ever.

Shaking these dark thoughts from his head, Halley circled the fringes of a crowd gathered around a large bonfire. The heat on his face was in stark contrast to the frosty air that chilled the rest of his body.

Paying little attention to the path in front of him, he almost collided with a gigantic man backing out of a doorway. Clad in heavy academic gowns from head to toe, the oversized man turned away, head down.

'And I say that you, sir, are a charlatan!'

The voice stopped Halley in his tracks. *Robert Hooke?*

In the doorway stood a hunched figure rendered comically small beside his giant of a companion. The crabby expression, narrowed eyes and sharp protruding nose were unmistakeable.

'Robert! It's me,' said Halley.

Hooke squinted.

'Edmond Halley. I'm back in England.'

Hooke's eyes widened. 'And not a moment too soon. This gentleman here was . . .'

The giant stepped forward. 'I am John Goad of Oxford.' He bowed with such a melodramatic flourish of his gowns that, if not for his bulk, he might have taken flight.

'Don't start all that again,' said Hooke, flapping away the billowing fabric.

'Mister Halley,' said Goad, 'I wish to address The Royal Society of London for Improving Natural Knowledge.'

'Then Mister Hooke is your man,' said Halley.

Goad and Hooke scowled at each other.

'He's the secretary and curator of experiments . . .'

'Also the Gresham Professor of Geometry and Surveyor to the city of London but you, sir, are a peddler of nonsense,' said Hooke emphatically to Goad. 'The Royal Society only welcomes gentlemen inclined to the experimental sciences.'

'I practise the oldest of the sky arts.'

Halley sighed. An astrologer . . .

'We have rejected astrology as false,' said Hooke.

'But I can bring personal testimonies of the accuracy of my horoscopes . . .'

'One needs only to sing a madrigal every day for a fortnight, and lo! it will cure you of your fever,' said Hooke. 'Is that to be trusted? Of course not. Without a clear understanding of how one thing links to another, you can believe in nothing.'

'I am aware that my art is falling from favour, but God could not have created the heavens without purpose. And that purpose is the weather – why does it rain one day and shine the next? Why? Because of the planets! The air around us is the shoreline with the heavens. It must be influenced by the planetary positions.' Goad looked from Halley to Hooke and back again. 'The movement of the planets is the engine of our seasons – God's engine for raising our crops, providing the water and the sunlight needed for them to grow.'

Hooke tutted.

Goad smiled indulgently. 'I understand your scepticism but consider this: Johannes Kepler, that paragon of astronomical achievement and the architect of the Sun-centred astronomy, was also a master astrologer.'

Halley said, 'Mister Goad, Kepler's laws of planetary motion can be tested telescopically and shown to be true. We reject the rest of his work. Astrology cannot be measured by any instrument that I know.'

'Do you similarly reject the Reverend Flamsteed? He cast a horoscope for the founding of the Greenwich observatory.'

'Have you seen that horoscope?' challenged Halley. 'I was there on the very day it was cast.'

Goad's face fell. 'You know the King's Astronomer?'

'I do, and written at the bottom of the page in John's own hand is *risum teneatis, amici – can you help laughing, friends*? He cast his horoscope as tradition dictated, not because he believed in it!'

Goad opened his arms. 'My friend, astronomers can tell us where the stars are, but it is up to astrologers to interpret meaning. Without us, the study of the heavens is futile.'

'The study of the heavens will allow us to safely navigate the oceans. I hardly call that futile.' Halley pointedly pulled his jacket shut.

'We do not believe in strange planetary forces communicating themselves across space,' added Hooke. 'That is magic. We believe only what we can measure.'

Goad flung his hands in the air. 'Measurement! Always measurement!'

'The measurement of nature is the only sure way of leading us back to God. Yes, Mister Goad, when Adam looked into a drop of water, he saw the microscopic life within it just as plainly as I have seen it using my microscope. When he looked into the sky, he saw all the stars in heaven and in so doing, he saw God. But, during the Fall, man's senses crumbled, trapping us in these tiny brains with these limited senses. So now we have to build telescopes and microscopes to rediscover the knowledge of creation, and to rediscover God.'

Goad looked aghast at Hooke, who ploughed on. 'Once our investigations have revealed nature's laws, we will be left with a collection of unexplainable phenomena that must therefore require God's direct intervention.'

Goad spluttered, 'Sounds like blasphemy to me. I tell you this with great confidence: my system of astrological weather prognostication will prove more important than any of your efforts at weighing the air.' He flapped his gowns.

'The term is *air pressure!*' called Hooke, bunching his fists and thrusting them onto his hips. 'And the barometer is an important scientific instrument!'

'Useless!' barked Goad, leaning forward threateningly.

Halley placed a hand on each man's chest. 'Enough! Let him go, Robert, there are those who will never understand.'

Goad growled but took off into the night. The bonfire crackled. A hawker bellowed his sales pitch for roasted chestnuts. A series of sharp reports split the air, and the two men looked up to see the brilliant trails of rockets shooting through the sky and blossoming into colour high above. Their attention was drawn back to Earth by a noisy gang of young apprentices who swaggered past them, swearing and shouting at onlookers to make way. To the men's horror, the lads appeared to be dragging a body.

'Make way for his Holiness!' the gang shouted.

No, not a man, Halley realised, but an effigy dressed in the white robes and mitre of the Pope.

Egged on by the crowd, they hefted the straw man onto a pyre. Bawdy cheers erupted and drowned the roar of the flames.

'When did England become so mean-spirited?' asked Halley.

'About the same time it dawned on people that the King was inching us back to Rome.'

'Do you really think Charles is a Catholic?'

'Of course, his brother has converted openly.'

'The Duke of York?'

Hooke nodded. 'Parliament has forced him to step down as Lord High Admiral.'

Halley could scarcely believe what he was hearing. Had England changed so much in the two years he had been away? 'But he's heir to the throne . . .'

'They're trying to exclude him from that, too. And, if they fail, we'll surely have a Catholic monarch – if we don't have one already . . . Tell me, how long have you been back?'

Halley inhaled the smoky night air. 'Some weeks, well, maybe a month or so.'

'And now you want to see me?'

'I've been busy making the final calculations for the star chart. I have to present it to the King next week, and it has to be perfect.'

'And you couldn't even drop in to say hello?' Hooke eyed him expectantly.

'Robert, I'm not the only thing that's back. So is the comet.'

'What?'

'The comet of last month has returned.'

'I haven't seen it.'

'It's not in the evening sky any more. It's appearing before dawn. It's as if it fell towards the Sun, swung round behind it and now travels outwards into space again.'

Hooke's eyes glittered in the firelight. 'Can we see it?'

'If we wait the night out and if you still have your telescope . . .'

It was nearly dawn before the bonfires died away into tiny, smoky columns dotting the city's skyline. The dampness in the air meant that it would be a race to see the comet before the soot combined with the morning mist to choke the city in a filthy brown miasma.

Halley scanned the horizon from the cramped roof platform that Hooke had commissioned atop Gresham College especially for his astronomical work. Hooke stifled a yawn and pulled out a handkerchief from his leather money pouch to wipe the telescope's lens and tube. The slim telescope stood ready between them.

Then Halley saw it, just a glimpse at first – a ghostly fan of light, no bigger than a thumb's width at arm's length. It was hanging above the rooftops, so faint that it danced in and out of visibility.

'There it is,' said Halley, pointing into the sky.

Hooke peered into the darkness. 'You're imagining it . . .'

Halley slid away to swing the telescope into position and duck his head to the eyepiece. Beautiful! The comet's tail reminded him of the long hairs that trailed a galloping horse, frozen in a portrait. He traced the threads of ethereal light to the head of the comet, where a small jewel glittered. 'No, come and look.'

Hooke bent to the eyepiece, knocking the tripod and forcing Halley to realign things. 'Well, I'll be . . .'

'Flamsteed thinks that it must be magnetic and that it's being repelled by the magnetism of the Sun,' said Halley.

'Impossible! If you melt a magnet it loses its magnetism, and the Sun must surely be molten to be so hot. Even Newton agrees with me on this.'

'You talk to Newton? I must surely have sailed back to some strange country and mistaken it for England.'

Hooke looked round from the eyepiece and straightened up. 'Not even I can hold a grudge for ever. We've been exchanging letters.'

'The controversy over the origin of colours is forgotten?' Halley moved closer, unable to keep the disbelief from his voice.

'Of course. I asked him if he could use his new mathematics to calculate why the planets stay in orbit.'

'To prove why Kepler's laws work?'

'Precisely.'

Halley was impressed. 'I admit I don't know much about his new mathematics. What's it called? Fluxions?'

'Fluxions and fluents,' Hooke said. 'No one knows much about them. He prefers to keep the method a secret, though I believe there are some papers lodged in the Society's archives, and Collins once told me that Newton had exchanged something with Leibniz in Hanover – apparently, he was thinking along similar lines.'

'Leibniz was?'

'Yes. Both working on a way to calculate the rates of change of moving quantities . . . but you're sidetracking me. Newton isn't really interested in Kepler's laws any more. He says he's abandoned natural philosophy altogether *and* he made a number of mistakes in his letters that I corrected for him.'

Halley registered the satisfaction in Hooke's voice. 'So, what is Newton doing these days?'

'Alchemy,' said Hooke.

'But that's illegal. You must be mistaken.'

'From what I hear, he hardly ever leaves his furnace.'

'I didn't have Newton down as a dabbler,' Halley said.

'He's consumed by some notion of concocting the Philosopher's Stone.' Hooke suddenly looked impish. 'Perhaps if he's thrown in gaol, he'll have some time to perform our calculations for us.'

'Perish the thought,' Halley said pointedly, making Hooke chuckle.

The unexpected sound of someone climbing the rickety staircase caught Halley's ears. He turned and locked eyes with a young woman.

'Mister Halley,' she said, her eyes smiling at him even if her mouth stayed level.

'Mistress Hooke, I'm delighted to see you again.'

'Will you be staying for breakfast?'

'He's not staying. He's just leaving. Off you go.' Hooke bustled round, barring her entry to the platform.

The woman allowed her gaze to linger on Halley even as she complied with her uncle's wishes and descended the staircase. Her image seemed to linger in the air where she had been standing.

'Grace has grown somewhat since last I saw her,' he said.

'Keep well away from her, Edmond.'

Hooke's sharp tone confused the young astronomer and he mumbled, not entirely truthfully, 'I remark on her as I would a spring flower.'

Hooke folded his arms. 'If you go now, you can still get a few hours' sleep before the day starts properly.'

'As you wish.' Halley knew better than to argue. He made his way to the stairs; the handrail was slick with dew. Two steps down, he stopped. 'Robert, just one thing. What you said to Goad, about finding God in the things we cannot understand. Did you really mean that?'

Hooke gave him an exasperated look. 'Of course.'

'But what happens if we can explain everything? I mean what if all our experimenting, all our observations and all our measuring give us a set of mathematical laws that can explain everything. What will we need God for then?'

'That will never happen.'

But what if it does? Halley thought. *What if it does happen?*

The author

Journalist, author and broadcaster, Stuart Clark has devoted his career to presenting the dynamic world of astronomy to the general public. He is a Fellow of the Royal Astronomical Society, a Visiting Fellow of the University of Hertfordshire and a former Vice Chair of the Association of British Science Writers. In 2000, UK daily newspaper *The Independent* placed him alongside Stephen Hawking and the Astronomer Royal, Professor Sir Martin Rees, as one of the 'stars' of British astrophysics teaching.

He divides his time between writing books and writing for the European Space Agency in his capacity as senior editor for space science, alongside producing features for magazines and newspapers. He has written seventeen books to date, selling more than 250,000 copies worldwide, which have been translated into twelve languages so far. He regularly lectures throughout the UK and, increasingly, around the world.

www.stuartclark.com